RED CELL STRUCTURE AND METABOLISM

Proceedings of a Colloquium
27 to 29 August 1969, Jerusalem, Israel

Edited by

BRACHA RAMOT, M.D.

Associate Professor of Medicine, Tel Aviv University Medical School
and Head of Section of Hematology, Tel Aviv University
Faculty of Continuing Medical Education, Tel Aviv, Israel

Academic Press   New York / London

Copyright © 1971, by ISRAEL JOURNAL OF MEDICAL SCIENCES, Jerusalem

All rights reserved. No part of this book may be reproduced in any form, by photostat, microfilm, retrieval system, or any other means, without written permission from the publishers.

Produced exclusively for ACADEMIC PRESS, INC., by
ISRAEL JOURNAL OF MEDICAL SCIENCES, Jerusalem

ACADEMIC PRESS, INC.
111 Fifth Avenue
New York, New York 10003

Distributed in the United Kingdom by
ACADEMIC PRESS, INC. (LONDON) LTD.
Berkeley Square House, London WIX 6BA

Library of Congress Catalog Card No. 70-142183
ISBN 0-12-577150-9

Printed in the United States of America

RED CELL STRUCTURE
AND METABOLISM

CONTENTS

Foreword   7

SESSION I

From chemical mechanism to clinical application   E. M. Kosower   11

The significance of intracellular glutathione   N. S. Kosower and E. M. Kosower   16

The sequestration of old red cells and extruded erythroid nuclei   D. Danon, Y. Marikovsky and E. Skutelsky   23

Red cell metabolism in heat-acclimated golden hamsters   N. Meyerstein and Y. Cassuto   39

Heinz body formation in red cells of the newborn infant   E. Kleihauer, A. Bernau, K. Betke and M. Keller   45

Discussion   51

SESSION II

Aspects of the structure, synthesis and clinical effects of unstable hemoglobins   R. F. Rieder   65

Hemoglobins with altered oxygen affinity   H. M. Ranney   80

Hemoglobin Torino disease. Clinical and biochemical findings   E. Gallo, G. Ricco, V. Prato, G. Bianco and U. Mazza   86

Properties of oxidized hemoglobin subunits separated *in vitro* and their relation to intraerythrocytic inclusion bodies in thalassemia   E. A. Rachmilewitz   94

Discussion   103

SESSION III

Diaphorases of human erythrocytes   E. Hegesh, N Calmanovici, M. Lupo and R. Bochkowsky   113

Studies on NADH and NADPH diaphorases in blood cells from normal subjects and in congenital methemoglobinemia   J. C. Kaplan   125

Electrophoretic and kinetic characterization of NADH-diaphorase variant in a methemoglobinemic subject J. M. Schwartz, J. M. Ross, P. S. Paress, K. Fagelman and L. Fogel 135

Discussion 147

SESSION IV

Some biochemical parameters of differentiation and maturation of synchronous erythroid cell populations Ch. Hershko, A. Karsai, L. Eylon, R. Schwartz and G. Izak 151

Red cell metabolism in iron deficiency anemia U. Mazza, G. P. Pescarmona, G. Bianco, R. Ricco and E. Gallo 164

The red cell nucleoside monophosphate kinases and their regulatory role in purine nucleotide metabolism A. Hershko and J. Mager 171

Evidence for an abnormal red cell population in refractory sideroblastic anemia I. Ben-Bassat, F. Brok-Simoni and B. Ramot 181

Discussion 189

SESSION V

Kinetic properties of pyruvate kinase and problems of therapy in different types of pyruvate kinase deficiency D. Busch, R. W. Hoffbauer, K. G. Blume and G. W. Loehr 193

New glucose-6-phosphate dehydrogenase variants in Israel. Association with congenital nonspherocytic hemolytic disease B. Ramot, I. Ben-Bassat and M. Shchory 206

Glycogen metabolism and glycolysis in erythrocytes from patients with glycogen storage disease type III and normal subjects S. W. Moses, N. Bashan and R. Chayoth 213

Discussion 226

FOREWORD

The past decade has witnessed a formidable escalation of scientific discovery in various aspects of Hematology and the transmission of this new knowledge has become a correspondingly difficult task. The interdisciplinary symposium with the participation of not only hematologists, but also investigators in various other disciplines, has proved an effective means of overcoming this difficulty. This concept provided the motivation for the present symposium, and hopefully for additional future symposia.

On this occasion, the deliberations were limited to certain aspects of red cell structure and function in which there have been major developments in the last few years. The discussion included contributions by investigators from various countries in the fields of organic chemistry, biochemistry and clinical medicine. It is hoped that many of the new and challenging ideas that evolved will give impetus to further research.

The Editor is indebted to all the contributors and discussants for the good will and dedication which characterized their contributions.

The symposium was organized by the Hematology Section of the Postgraduate Medical School of Tel Aviv University and the Israel Society of Hematology and Blood Transfusion. Special thanks are due to the Israel Journal of Medical Sciences and its Editor-in-Chief, Dr. Moshe Prywes, for his initiative in publishing this volume and to Editorial Secretaries, Rebekah Soifer and Audrey Young for their help in its compilation.

SESSION I

Chairmen: R. F. Rieder, *USA*
            B. Ramot, *Israel*

Participants: A. Bernau, *West Germany*
              K. Betke, *West Germany*
              Y. Cassuto, *Israel*
              D. Danon, *Israel*
              M. Keller, *West Germany*
              E. Kleihauer, *West Germany*
              E. M. Kosower, *USA*
              N. S. Kosower, *USA*
              Y. Marikovsky, *Israel*
              N. Meyerstein, *Israel*
              E. Skutelsky, *Israel*

FROM CHEMICAL MECHANISM TO CLINICAL APPLICATION

EDWARD M. KOSOWER

Department of Chemistry, State University of New York at Stony Brook, Stony Brook, New York, USA

The role of chemical mechanism in arriving at clinical applications is discussed for organophosphate intoxication, acute intermittent porphyria, sickle cell anemia and glutathione dyscrasias.

The road from fundamental discovery to practical application is often long and it is especially difficult when an effective clinical agent for humans is desired. Therapeutic agents have almost always been found as a result of accidental discoveries, e.g. antibiotics: penicillin; insecticidal agents: DDT; antibacterial agents: sulfonamides; and developed along empirical lines (acetylsalicylic acid, antifolic acid agents). The success of this effort is well known. However, if we measure the current rate of progress against the complexity of the medical problems which face us today (drug-resistant bacteria, cancer, atherosclerosis, heart disease, individualized treatments required because of genetic differences, etc.), we sense the inadequacy of purely empirical methods. My purpose today is to take up a number of medical problems: organophosphate intoxication, acute intermittent porphyria, sickle cell anemia, and glucose-6-phosphate dehydrogenase deficiency; and to discuss molecular approaches to their elucidation and treatment (actual or potential). The fact that three out of the four problems are genetic diseases is understandable since a powerful technique for probing a normal system involves the study of the changes which ensue upon alteration of a key component in the system.

The most important molecular alteration introduced into living organisms by intoxication with most organophosphates is phosphorylation of a serine hydroxyl group in the enzyme acetylcholinesterase (AChase) (equation 1):

$$\text{AChase OH} + (\text{i-PrO})_2\text{P(O)F} \rightarrow \text{AChase OP(O) (Oi-Pr)}_2 \text{ inhibited enzyme} \quad (1)$$

Normally the enzyme functions at synapses (especially neuromuscular junctions) to hydrolyze the acetylcholine (ACh) released as a transmitter, permitting recovery of the nerve and making possible subsequent depolarization by another dose of ACh (1) (equation 2):

$$\text{ACh} + \text{receptor} \rightleftharpoons \text{ACh} \cdot \text{receptor complex} \rightarrow \text{neurone firing} \quad (2)$$

$$\bigg| \text{AChase}$$

$$\longrightarrow \text{choline} + \text{acetate}$$

It was noted by Nachmansohn and Wilson that inhibited enzyme (equation 1) spontaneously recovered activity on standing in water for a long time. A search was made for nucleophiles which could accelerate the hydrolysis of the phosphorylated enzyme (equations 3 and 4):

$$H_2O + \text{inhibited enzyme} \rightarrow \text{active enzyme} + \text{phosphoric acid derivative} \quad (3)$$

$$\text{Nu} + \text{inhibited enzyme} \rightarrow \text{active enzyme} + \text{phosphoric acid derivative of nucleophile} \quad (4)$$

Wilson designed a successful molecule on the basis that a) oximes were highly reactive towards phosphoric acid esters and b) a cationic center had to be present in the molecule at a distance from the ester group similar to the distance between the trimethylamino group and the acetate group in ACh (I). The agent designed on these physical organic principles was PAM (II) 1-methyl-2-aldoximopyridinium iodide, and it, or closely related agents, is in widespread use in the treatment of organophosphate (e.g., parathion) poisoning.

A genetic disease, acute intermittent porphyria (AIP) (2) is common in the Republic of South Africa and occurs infrequently elsewhere (1.5 cases/100,000 population). Pain, paralysis or other neurological symptoms several days after barbiturate administration often bring AIP to mind as a possible diagnosis. The disease is difficult to manage, rather unpredictable, and frequently fatal.

The genetic lesion in AIP is unknown, but the biochemical effect is clear: many compounds which normally stimulate an increase in the "cytochrome P-450" system and thereby enhance their own metabolic disposal, lead to an abnormal increase in the level of δ-aminolevulinic acid (ALA) synthetase and thus in the production of ALA from succinoyl-CoA and glycine. The subsequent dehydration of ALA to porphobilinogen (PBG) raises the levels of PBG.

We reasoned that the neurological symptoms in AIP arose from the excess ALA being transported to the brain as a pyridoxal-Schiff base derivative (III). In the brain, this derivative replaced the derivative of the transmitter γ-aminobutyric acid (GABA) (IV) (3).

We therefore synthesized a diazoester ($N_2CHCOCH_2CH_2COOCH_3$) which we hoped would permanently inhibit ALA synthesis and therefore ameliorate the condition of AIP sufferers. In fact, this diazoester abolished the production of PBG completely and affected ALA levels only to a limited extent. We were able to show that the neurological condition of rats with chemically-induced porphyria (CIP) and AIP, did not change with diazoester administration. We thus feel that we have good evidence that our original hypothesis is valid: excess ALA is responsible for the symptomatology and that an inhibitor of the synthetase would be a useful therapeutic agent for AIP. The reasoning behind the choice of diazoester is outlined elsewhere (3), and is based on physical organic principles.

Sickle cell anemia is a genetic disease arising from the substitution of a valine for a glutamic acid residue at the sixth residue of the β-chain in hemoglobin. The hydrophobic nature of the valine raises the probability of intermolecular interaction; deoxyhemoglobin S ($\alpha_2\ \beta_2\ ^{6\ glu} \rightarrow$ val) separates from solution within the red blood cell and causes the cell to assume its characteristic sickle shape.

We have modified hemoglobin S with glutamic acid residues by reaction with the N-carboxyanhydride of glutamic acid. Some increase in solubility is noted, even with small degrees of modification (4). Although such treatment seems to be a rather drastic therapeutic measure, it is no worse than some which have been suggested for treatment of sickle cell anemia: change of blood pH, change of blood temperature, use of high oxygen pressure, use of atmosphere of propane, etc. A practical therapeutic agent (one which will penetrate the cell and react more or less specifically with hemoglobin) is not yet known. The principle behind its design is clear, however. The modification should produce a hemoglobin molecule with additional charged groups, lowering in this way the chance for intermolecular interaction.

We have referred to the process of chemical modification of proteins as "protein transformation" (5), and believe that the future will bring us some results of fundamental and clinical importance.

The last topic I wish to mention is that of intracellular glutathione (GSH). The high GSH content of red blood cells is well known and the overall consequences of the failure of the cell to maintain the GSH concentration because of glucose-6-phosphate dehydrogenase deficiencies are serious. Our findings in this area will be discussed by Dr. N. Kosower in the following paper. Of concern to us at this moment is how we came to discover reagents for the intracellular oxidation of GSH to oxidized glutathione (GSSG). Hematologists had for many years used acetylphenylhydrazine to "oxidize" intracellular GSH without paying particular attention to the requirement for oxygen in the reaction (6) (equation 5):

$$GSH + ?\ C_6H_5NHNHCOCH_3 \xrightarrow{O_2} GSSG + ? \qquad (5)$$

Any chemist would find the role of acetylphenylhydrazine as an oxidizing agent anomalous—it is much more likely that it is a reducing agent. We therefore postulated that oxygen converted APH into the true GSH agent, acetylphenyldiazene (equation 6):

$$C_6H_5NHNHCOCH_3 \xrightarrow{O_2} C_6H_5N=NCOCH_3 \qquad (6)$$

We then tried a closely related compound, methyl phenyldiazenecarboxylate (azoester), and discovered that GSH was oxidized to GSSG stoichiometrically (equation 7):

$$2GSH + C_6H_5N=NCOOCH_3 \rightarrow GSSG + C_6H_5NHNHCOOCH_3 \quad (7)$$

We have worked out the mechanism of the reaction and have utilized this compound and others on red blood cells and many other biological systems (7,8). Potential applications have been discussed elsewhere (7):

In closing, let me remind you that I have given illustrations (of necessity, without proper detail) to indicate how a detailed knowledge of chemical mechanism can lead to results of clinical importance. If the latter goal has not been achieved in all cases, you will understand the difficulties. We feel that our approach will be useful in many other medical problems.

The author is grateful to the National Science Foundation for a Senior Postdoctoral Fellowship during 1968–1969, making it possible for him to spend the year at the Department of Biophysics, Weizmann Institute, Rehovot, Israel.

REFERENCES

1. NACHMANSOHN, D. "The molecular basis of nervous activity." New York, Academic Press, 1959.
2. KOSOWER, N. S. and LONDON, I. M. The porphyrias, in: Barnett, H. L. (Ed.), "Pediatrics," 14th edn. New York, Appleton-Century-Crofts, 1969.
3. KOSOWER, N. S., KOSOWER, E. M., ZINN, A. B. and CARRAWAY, R. Methyl-5-diazolevulinate intervention in chemically-induced porphyria of rats. *Biochem. Med* **2**: 389, 1969.
4. KOSOWER, E. M., KOSOWER, N. S. and LACOURSE, P. C. The solubilization of deoxyhemoglobin S. *Proc. nat. Acad. Sci. (Wash.)* **57**: 39, 1967.
5. KOSOWER, E. M. The therapeutic possibilities arising from the chemical modification of proteins. *Proc. nat. Acad. Sci. (Wash.)* **53**: 897, 1965.
6. BEUTLER, E., ROBSON, M. and BUTTENWIESER, E. *J. clin. Invest.* **36**: 617, 1957.
7. KOSOWER, E. M. and KOSOWER, N. S. Lest I forget thee, glutathione... *Nature (Lond.)* **224**: 117, 1969.
8. KOSOWER, N. S., SONG, K. R., KOSOWER, E. M. and CORREA, W. Glutathione I, II, III, IV. *Biochim. biophys. Acta (Amst.)* **192**: 1, 1969.

THE SIGNIFICANCE OF INTRACELLULAR GLUTATHIONE*

NECHAMA S. KOSOWER** and EDWARD M. KOSOWER***

Department of Medicine, Albert Einstein College of Medicine, Bronx, New York and Department of Chemistry, State University of New York at Stony Brook, New York, USA.

Glutathione (GSH) is a universal constituent of biological systems. Progress in understanding the roles of GSH has been hampered by the lack of a convenient method for temporary alteration of its concentration within the intact cell or organism in a reasonably specific way. E. M. Kosower, in the preceding paper, outlined the principles that led to the development of a new class of reagents which are suitable for this purpose.

These compounds, which have the general formula RN = NCOX or XCON = NCOX, readily penetrate the cell. They cause a rapid oxidation of GSH to GSSG, even at low temperature, at which the metabolic activity of the cell is minimal. The reaction is stoichiometric, permitting control of the extent of GSH oxidation. The reaction is oxygen-independent and, therefore, the GSH can be oxidized in cells under anaerobic conditions. The capacity of the system for the reduction of GSSG to GSH is retained. Thus, the effects of a temporary loss of GSH on the functioning of cells, tissues or whole organisms can be evaluated (1–3).

The choice of reagent is made according to the requirement of a particular study. Agents which are readily hydrolyzed rapidly disappear from the system to which they are applied. They are therefore short term thiol oxidizing agents. In addition, compounds which produce free radicals

---

* This work was carried out in part at the Weizmann Institute of Science, Rehovot, Israel, during sabbatical leave, in collaboration with David Danon and Yehuda Marikovsky.
** Recipient of Career Development Award (1–K4–GM–38889) from the National Institute of General Medical Sciences, U.S. Public Health Service.
*** A National Science Foundation Senior Postdoctoral Fellow for 1968–1969 at the Department of Biophysics, Weizmann Institute of Science, Rehovot, Israel.

through hydrolysis and reaction with oxygen provide intracellular chemical challenge to the system. Other agents are, in contrast, stable towards hydrolysis and thus are long term thiol oxidizing agents, and can be used to maintain a low intracellular GSH concentration for a considerable time (1, 4).

In the present paper, we will concentrate on studies done on red blood cells, and describe the effects of two different reagents on the red cell. These will serve to illustrate the characteristics of the GSH oxidation reaction and some of the consequences of a temporary absence of GSH.

The first compound is methyl phenyldiazenecarboxylate, azoester, $C_6H_5N=NCOOCH_3$. The chemical properties of azoester are shown in Fig. 1; in common with all other reagents, it oxidizes GSH to GSSG. A fraction of azoester is consumed by protein amino groups to give a biologically unimportant byproduct. An important property of azoester is that it is unstable and can give rise to free radicals, through hydrolysis and reaction with oxygen, when GSH is exhausted.

When human erythrocytes are mixed with appropriate amounts of azoester, part or all of the GSH is oxidized within 1 to 2 min. The reduction of GSSG to GSH within the normal human erythrocyte is achieved by

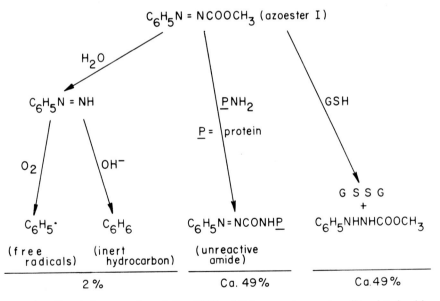

FIG. 1. Chemical properties of the GSH-oxidizing agent azoester. (Reprinted with permission from *Biochem. biophys. Res. Commun.* **37**: 593, 1969, copyright by Academic Press.)

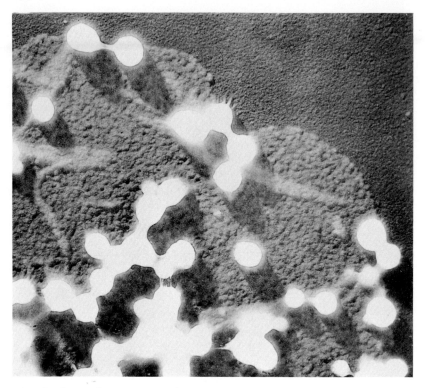

FIG. 2. A ghost of azoester-treated oxygenated erythrocyte. Multiple electron dense masses are seen. The membrane structure is preserved, as shown by the "granular" appearance of the membrane. × 30,000.

subsequent incubation of the treated cells in the presence of glucose (2, 3, 5). If azoester is used in amounts insufficient for oxidation of most of the cellular GSH, no damage to the cell is observed upon subsequent incubation. GSH regeneration from GSSG is almost complete, no intracellular protein denaturation occurs and no membrane injury is found. Treatment of red blood cells with reagent in excess of the amount reacting with GSH permits the free radicals which are formed to cause cellular damage to the GSH depleted cell. In the erythrocyte containing oxyhemoglobin, free radicals are formed throughout the cell and lead to intracellular protein denaturation: upon incubation of oxygenated erythrocytes after treatment with excess azoester, multiple Heinz bodies are formed, shown in Fig. 2 as electron dense masses. The significant intracellular damage is accompanied by only a mild effect on the membrane, which is manifested as a diminution in surface charge. Incubation of erythrocytes rendered poor in intracellular oxygen (by loading them with carbon monoxide) and suspended in oxygenated buffer after treatment with azoester leads to hemolysis *in vitro* (6). Hemolysis is only observed after the use of quantities of azoester greater than that required to oxidize all the intracellular GSH to GSSG (Fig. 3). Under these conditions, the free radicals are generated from excess azoester at or near the membrane, leading to severe membrane damage. The membrane damage occurs in the absence of any significant intracellular protein denaturation (Fig. 4).

This model for *in vitro* lysis is applicable to tissue cells, since most cells are oxygen-poor. Indeed, treatment of many different cell types with excess azoester leads to lysis of these cells.

A second compound used in these studies is diamide $(CH_3)_2$ NCON = NCON $(CH_3)_2$. The oxidation of GSH by diamide follows the general

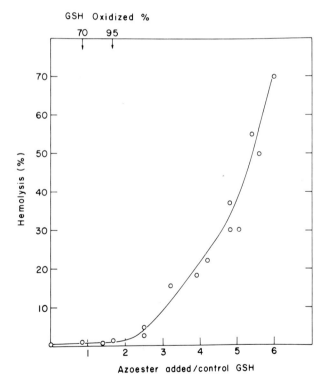

FIG. 3. Hemolysis of azoester-treated oxygen-poor cells. (Reprinted with permission from *Biochim. biophys. Acta* (*Amst.*) **192**: 23, 1969.)

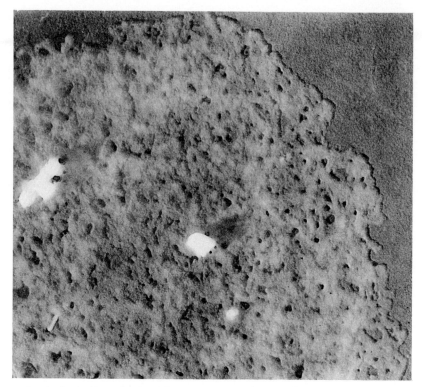

FIG. 4. A ghost of azoester-treated oxygen-poor erythrocyte. Complete loss of membrane structure, irregular edges and multiple holes are seen. × 30,000.

characteristics described above. Special properties of this compound include stability towards hydrolysis and no tendency to form free radicals. In oxygen-rich red cells treated with excess diamide, Heinz body formation does not occur upon incubation. No lysis is observed when oxygen-poor cells are incubated after treatment with excess diamide. GSH regeneration is delayed according to the amount of diamide used.

In comparing the effects of these two compounds on the red blood cell, we arrive at the following conclusion: No damage occurs unless most of the GSH within the erythrocyte is oxidized and a chemical challenge is present during the time when the concentration of GSH is low. Thus, treatment of cells with azoester in amounts insufficient to oxidize most of the intracellular GSH has little observable effect. Thiol oxidizing agents which do not generate free radicals cause no damage even when used in excess, leading to depletion of GSH for some length of time. When GSH is temporarily absent, certain additional challenge, as an excess reagent generating free radicals, would lead to damage.

The ultimate destruction of the erythrocyte is associated with membrane failure. Azoester-treated oxygenated and carbon-monoxide loaded erythrocytes can serve as models for the behavior of red cells challenged with an oxidizing agent: a subtle alteration in red cell surface properties can be produced even though most of the damage is intracellular. Through alterations in surface charge, the cell so changed might be "recognized" by the sequestration system and taken out of the circulation (7). Certain oxidizing agents might concentrate in the cell membrane as a result of their lipophilicity; the chemical challenge generated at or near the membrane might lead to severe membrane damage, manifested by intravascular lysis. In both cases, the damage resulting from chemical challenge at the time of GSH depletion occurs in spite of the subsequent regeneration of GSH in the cells. The function of cellular GSH in protecting membrane integrity against a chemical challenge is one of the roles of GSH in biological systems. Observed effects of these reagents on other cells or whole organisms have included, apart from injury to membranes, changes in a variety of metabolic functions of the cells, such as growth and division (1). In the mature erythrocyte, which lacks the apparatus for division or protein synthesis, a major role of GSH seems therefore to be the prevention of membrane injury by some reagents.

The results presented here are relevant to problems arising in some clinical situations: absence of red cell GSH or an inadequate ability to reduce GSSG to GSH is associated with a premature destruction of the erythrocyte, and severe hemolysis can result after exposure of susceptible individuals to certain drugs in moderate amounts. These same drugs can also cause hemolysis in normal individuals when used in high doses (8). Since the membrane injury, caused by challenge to GSH-depleted cells, is not prevented by any subsequent regeneration of GSH, it is important to aim at a regulation of drug dosage so as not to severely lower the cellular GSH at any time. Another possibility would be the modification of drugs to either diminish their GSH oxidizing capacity or to minimize the occurrence of reactive metabolites, in cases when one or the other of these two properties are not a necessary part of the therapeutic action of the drug.

REFERENCES

1. Kosower, E. M. and Kosower, N. S. Lest I forget thee, glutathione... *Nature (Lond.)* **224**: 117, 1969.
2. Kosower, N. S., Song, K. R. and Kosower, E. M. Glutathione, I. The methyl phenyldiazenecarboxylate (Azoester) procedure for intracellular oxidation. *Biochim. biophys. Acta (Amst.)* **192**: 1, 1969.
3. Kosower, N. S., Song, K. R. and Kosower, E. M. Glutathione III. Biological aspects of the azoester procedure for oxidation within the normal human erythrocyte. *Biochim. biophys. Acta (Amst.)* **192**: 15, 1969
4. Kosower, N. S., Kosower, E. M., Wertheim, B. and Correa, W. Diamide, a new reagent for the intracellular oxidation of glutathione to the disulfide. *Biochem. biophys. Res. Commun.* **37**: 593, 1969.
5. Kosower, N. S., Vanderhoff, G. A. and London, I. M. The regeneration of reduced glutathione in normal and glucose-6-phosphate dehydrogenase deficient human red blood cells. *Blood* **29**: 313, 1967.
6. Kosower, N. S., Song, K. R. and Kosower, E. M. Glutathione, IV. Intracellular oxidation and membrane injury. *Biochim. biophys. Acta (Amst.)* **192**: 23, 1969.
7. Danon, D., Marikovsky. Y. and Skutelsky, E. On the sequestration of old red cells and extruded erythroid nuclei, in: Ramot B. (Ed.), "Clinical aspects of red cell structure and metabolism." New York, Academic Press, 1970.
8. Beutler, E. Drug induced hemolytic anemia. *Pharmacol. Rev.* **21**: 73, 1969.

THE SEQUESTRATION OF OLD RED CELLS AND EXTRUDED ERYTHROID NUCLEI

D. DANON, Y. MARIKOVSKY and E. SKUTELSKY

Section of Biological Ultrastructure, Weizmann Institute of Science, Rehovot, Israel

Many studies have been carried out in recent years on red cell aging, most of them employing a biochemical approach. It was generally assumed that some prior biochemical insufficiency must be responsible for the death of the cell. An impressive number of searches have been made, in various laboratories, for one or another enzyme displaying a steep decline in activity towards the end of the physiological life span of the cell, and there are still groups looking for the "critical enzyme." While some of these studies have yielded important contributions to the understanding of red cell metabolism in health and disease, the fact that normal red cells, under physiological conditions, do not die or lyse while circulating in the blood vessels has been generally overlooked. The senescent cells are "taken out" of the circulation by the reticuloendothelial system (RES), mainly the spleen. For example, a cell on the day that it is 119 days old still circulates in the blood vessels, passes several times into the spleen, is held up in this organ for a short period and then reenters the circulatory system. On the so-called 120th day, in one of the passages of the cell through the spleen, a macrophage will establish contact with the old cell, phagocytize it and disintegrate it in its phagocytic vacuole. While our notion of "young" and "old" is based on a chronological statistical reference, the macrophage has criteria by which it can "recognize" which cell is old and should be removed from the circulation. The macrophage can also recognize damaged cells produced by various noxious agents.

The specific role of the macrophage is its recognition of a "nonself." The ability to recognize the surface characteristics of the antigenic map is relatively easy to imagine. It is much more difficult to imagine that the macrophage is able to titrate the level of enzymatic activity within the red cell and thus recognize it as young or old. It seems more reasonable to assume that changes in the biophysical properties of the membrane can be detected by the macrophage. Of all the biochemical and biophysical alterations (1-3) the surface charge, which is markedly reduced in older cells (4-6), appears to us the most likely feature to be recognized by the macrophage.

Separated fractions of "old" red cells move more slowly in an electric field than cells from the "young" fraction taken from the same blood sample (6). It has been demonstrated that the negative charge on the red cell surface can be ascribed almost entirely to the carboxylic group of neuraminic acid (7-9). N-acetyl neuraminic acid (NANA) was also found to be the major component in the reaction of the negatively charged red cell surface with the positively charged poly-L-lysine (10). A positive correlation was demonstrated between the rate of agglutinability with poly-L-lysine and the electric mobility of the cells (5, 11). The substrate for the receptor-destroying enzyme (RDE) on the red cell surface is neuraminic acid (12). Treatment with neuraminidase reduces the electric mobility of red cells (13). Adsorption of myxovirus, followed by elution, results in a marked decrease in the electric mobility of red cells (14) and the degree of reduction depends on the myxovirus type (15). Incubation of red blood cells with RDE derived from *Vibrio cholerae* (15) cultures almost completely removes the negative charge from the cells (5).

These features provide a system in which we can reduce the surface charge on the cell, thus facilitating attempts to establish a correlation between surface charge and survival. Another model system in which the macrophage recognizes an altered or deteriorated entity would be helpful in establishing whether a low surface charge on the membrane is a common feature in both recognition processes. Such a model is presented by the nucleus of the late erythroblast, which on expulsion is surrounded by a narrow rim of cytoplasm and membrane.

It has been pointed out that if the nucleus is always expelled, more free nuclei would be present in the circulating blood than are actually found (16). However, we can assume that the rarity of free nuclei in the hemopoietic centers may be ascribed to the avidity with which the macrophages phagocytize the expelled nuclei (17-23). It is interesting to query (22) in what respect the membrane surrounding the expelled nucleus differs from that which envelops the remaining future reticulocyte? What makes it "recognizable" by the macrophage? Direct measurement of the electric mobility or of the agglutinability of expelled nuclei is not feasible because the expelled nuclei are very rarely, if ever, found in the circulation and it is practically impossible to separate them from the bone marrow. We have therefore employed a

TABLE 1. *Electric mobility of young and old human red blood cells*

Blood sample no.	Young erythrocytes (top fraction)	Old erythrocytes (bottom fraction)
1	1.34	0.94
2	1.47	0.97
3	1.42	1.12
4	1.51	1.16
5	1.29	1.06
6	1.42	1.10
7	1.47	1.18
8	1.47	1.04
9	1.47	1.04
10	1.26	1.14
Mean	1.412	1.075

Electric mobility (μ/sec · v per cm) of separated young and old red cells from 10 different human blood samples. Electric mobilities of 30 cells in each direction were measured and the average of each blood sample calculated.

TABLE 2. *Correlation between electric mobility of red blood cells and the density of colloidal iron oxide particles counted on electron micrographs[a] for comparative evaluation of surface charge*

Blood sample	No. of particles/μ length of membrane		Electric mobility (μ/sec · v per cm)	
		Decrease in %		Decrease in %
Human red blood cells				
Young	20.4		1.32	
Old	12.6	35.0	1.03	24
10 units/ml RDE	11.8	38.0	0.96	28
20 units/ml RDE	0.0		0.27	80
Rabbit red blood cells				
Young	17.8		0.70	
Old	12.0	32.5	0.60	14
10 units/ml RDE	10.8	39.0	0.57	18
20 units/ml RDE				

[a] From each sample, the membranes of 25 cells were measured on the micrograph in the parts where the membrane is perpendicularly sectioned. The iron particles were counted on these parts.

system whereby we can analyze the differences in charge density on the various membrane surfaces using positively charged colloidal ferric oxide particles visualized on the membrane surface by electron microscopy, as

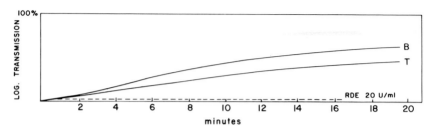

FIG. 1. The rate of agglutination of human red blood cells by poly-L-lysine, n=100. Agglutination curves of "young" red cells from a top fraction (T), "old" red cells from a bottom fraction (B) and a whole population of red cells treated with 20 units/ml RDE (– – – – –). Note that at any time after onset of agglutination the old cells are agglutinated in a rate approximately 30% higher than that of the young cells. No agglutination occurs after RDE treatment.

described by Gasic et al. (24). This approach has also allowed us to compare the surface charge on marrow and circulating cells.

The techniques for preparation of the red cells, their separation into age groups, their preparation for measurements of the agglutination kinetics, and the methods used for labeling of bone marrow and for counting the colloidal iron particles on micrographs were described by us in two recent papers (25, 26).

Table 1 shows the reduced electric mobility of "old" red blood cells as compared with cells of the "young" fraction. In these experiments (4), the cells were separated into "old" and "young" fractions by centrifugation according to Prankerd (27). A later technique for separating cells of different density by differential flotation (28) enabled us to separate red cells into age groups more efficiently. Using a battery of 20 separating phthalate ester mixtures of decreasing specific gravity (Gravikit, Miles Yeda, Rehovot, Israel), we were able to confirm our early data and show that there is a correlation between electric mobility and agglutination kinetics by the positively charged poly-L-lysine molecules (5,11). With both techniques, we could demonstrate that cells from the "old" fraction are less negatively charged than cells from the young fraction (Table 2 and Fig. 1). These data were confirmed by Yeari (6), who fractionated a red cell population into eight density groups (Fig. 2). Using the method of labeling the surface charge on cell membranes with colloidal iron, it can be seen that while the young red blood cells are densely labeled (Fig. 3), the old red cells are less labeled, showing patches devoid of colloidal iron particles (Fig. 4).

Treatment of young human cells with such amounts of neuraminidase

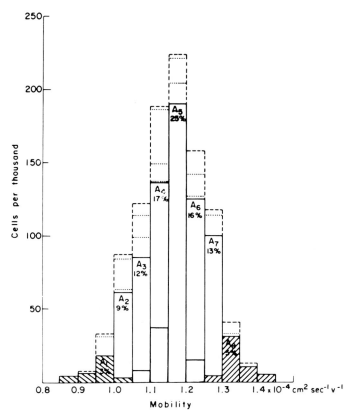

FIG. 2. Bar graph illustrating the mobility of cells in an electric field in the eight fractions separated according to their density. On the top of each bar, which constituted the peak of a fraction, its fraction number and the percentage it constituted of the whole population are given. Each bar indicates the % of cells that migrated at the speed noted in the abscissa. Since the various fractions had overlapping migration speeds, the additive effect from other fractions at each migration speed is shown on top of the column indicating the appropriate values in dotted lines. Fractions no. 1, 5 and 8 are shadowed to illustrate the distribution of migration speeds within these three representative fractions. (Reprinted, with permission, from ref. 6.)

and under such incubation conditions that about 30% of the surface charge is removed (25), shows that the electric mobility is reduced accordingly and the fragiligraph recordings of agglutinability by poly-L-lysine reveal a higher and more rapid deflection, like that of old cells (see Fig. 1). Electron microscopy reveals practically the same density and distribution of colloidal iron particles adsorbed to the cell membrane in those cells as in old cells

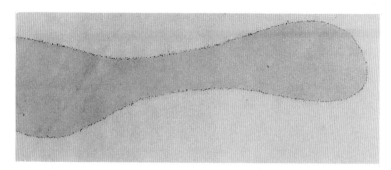

FIG. 3. Human red blood cells from a "young" fraction separated by differential flotation, fixed in glutaraldehyde and labeled with a nondialyzed positive colloidal iron suspension. Colloidal particles are uniformly distributed along the membrane surface. × 14,000. (Reprinted, with permission, from ref. 25.)

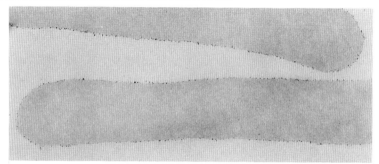

FIG. 4. Human red blood cells from an "old" fraction separated by differential flotation, fixed in glutaraldehyde, labeled with a nondialyzed positive colloidal iron suspension. Iron particles are deposited irregularly, leaving unlabeled gaps on the membrane surface. × 14,000. (Reprinted, with permission, from ref. 25.)

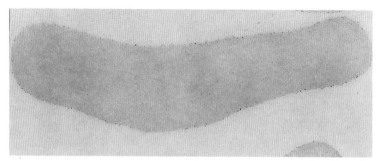

FIG. 5. Unseparated human red blood cells treated with RDE (10 units/ml) and labeled with nondialyzed positive colloidal suspension. Deposition of colloidal iron particles is similar in amount and disposition to that of an old human red blood cell. (See Fig. 4.) × 14,000. (Reprinted, with permission, from ref. 25.)

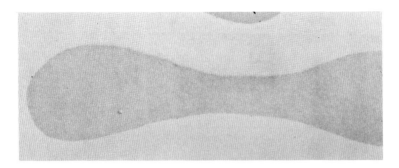

FIG. 6. Unseparated human red blood cells treated with RDE (20 units/ml) and labeled with nondialyzed positive colloidal iron suspension. There is practically no deposition of colloidal particles on the membrane surface. × 14,000. (Reprinted, with permission, from ref. 25.)

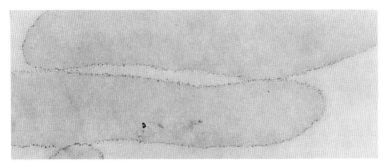

FIG. 7. Young rabbit red blood cells, labeled with a nondialyzed positive colloidal iron suspension, showing a deposition similar to that of young human red blood cells. × 14,000. (Reprinted, with permission, from ref. 25.)

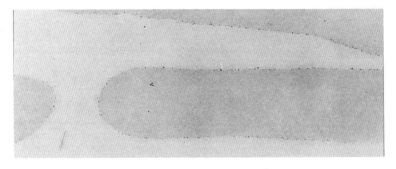

FIG. 8. Old rabbit red blood cells, labeled with a nondialyzed positive colloidal iron suspension, showing particles deposited irregularly, leaving unlabeled gaps on the membrane surface. × 14,000. (Reprinted, with permission, from ref. 25.)

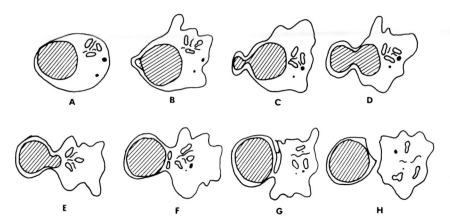

FIG. 9. Schematic representation (A to H) of the sequence of events occuring in the process of extrusion of the nucleus from the late erythroblast. (Reprinted, with permission, from ref. 22.)

(Fig. 5). If a more severe treatment with neuraminidase, using higher doses, is applied to the cells, practically no surface labeling can be detected (Fig. 6) and the fragiligraph recording of agglutinability by poly-L-lysine shows no deflection. In electric mobility measurements, only about 20% of the original mobility is observed. Fig. 7 and 8 show that similar differences exist between old and young rabbit erythrocytes.

Assuming that this reduction in surface charge demonstrated by three different methods is recognized by the macrophage, let us now look at what happens to the nucleus that is expelled from the late erythroblast and is also "recognized as an undesirable self" and phagocytized by the macrophage. The series of events that result in the expulsion of the nucleus, surrounded by a narrow rim of cytoplasm and plasma membrane are summarized in Fig. 9. This plasma membrane, or at least part of it, was, a few moments previously, a part of the erythroblast that was occasionally in close contact with the macrophage, but was not phagocytized. It was apparently recognized by the macrophage not only as a self, but as an undamaged desirable self. In the time lapse of a few minutes, a part of this late erythroblast is separated (Fig. 10) and, at this stage, it is approached by the macrophage and recognized as an "undesirable self" or "deteriorated self." The expelled nucleus is then phagocytized (Fig. 11). The question may be asked what is the difference between the membrane that envelops the nucleus, and the rest of the cell, which makes it "recognizable" by the

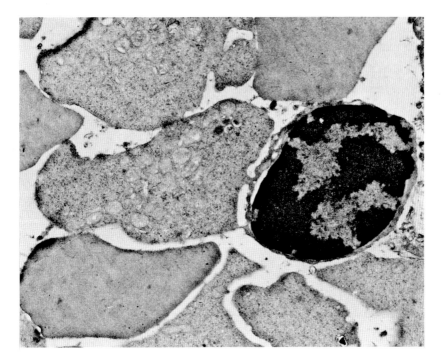

FIG.10. Electron micrograph showing the separation of the expelled nucleus and its surrounding plasma membrane from the remaining future reticulocyte. The nucleus, which has reassumed its typical rounded form, is now connected to the main bulk of cytoplasm by a wide bridge. Some vacuoles appear within this bridge. These openings become elongated, resembling the cleavage furrow which appears at the equatorial plane at the telephase of mitotic division of erythroblasts. The vacuoles continue to elongate parallel to the nuclear membrane until the extruded nucleus is attached to the cytoplasm by only one or two tiny connections. × 12,000. (Reprinted, with permission, from ref. 22.)

macrophage? Labeling with colloidal iron (Fig. 12, 13) suggests at least a partial answer to the question. The colloidal iron particles on the surface of the membrane that envelops the nucleus are less numerous than on the membrane of the future reticulocyte. This visual impression was confirmed by counting the number of particles/μ of perpendicularly sectioned membrane on micrographs at a magnification of × 30,000. It was thus shown that the density of colloidal iron particles on the membrane that envelops the expelled nucleus is only about 50% of that on the reticulocyte membrane. The macrophage apparently "recognizes" this reduced surface charge (Fig. 14).

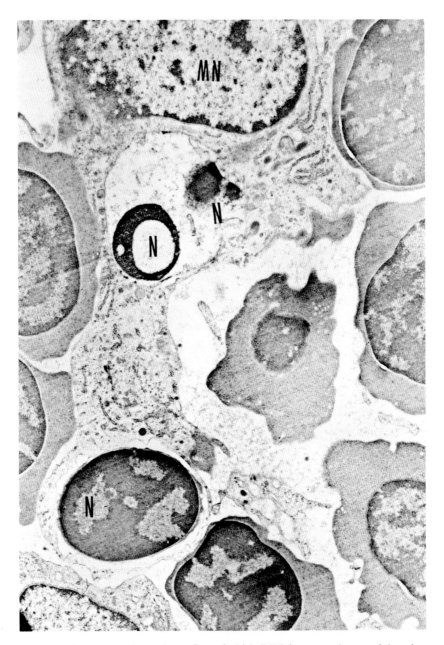

FIG. 11. A large macrophage, the nucleus of which (MN) is seen at the top of the micrograph, contains three extruded erythroid nuclei (N) at different degrees of disintegration. It is in close proximity to some late erythroblasts. × 10,000.

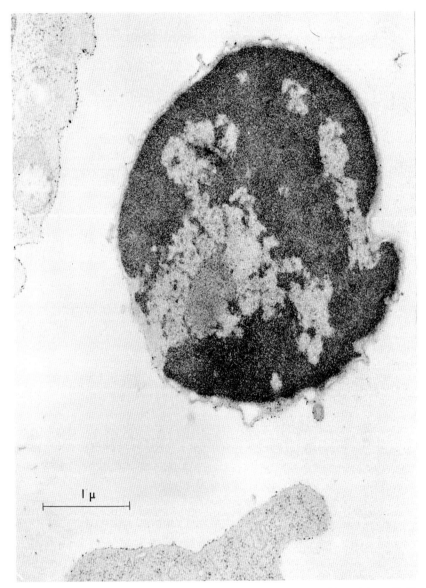

FIG. 12. Thin section of rabbit bone marrow suspension, stained with positive colloidal iron suspension. An extruded nucleus, surrounded by a narrow rim of cytoplasm and plasma membrane, is seen in the vicinity of erythroid (bottom) and leukoid elements. Heavy deposits of the colloid appear on the leukoid cell membrane as compared to those on the membrane of the erythroid cell. Almost no colloidal deposits appear on the free nucleus. × 23,000. (Reprinted, with permission, from ref. 26.)

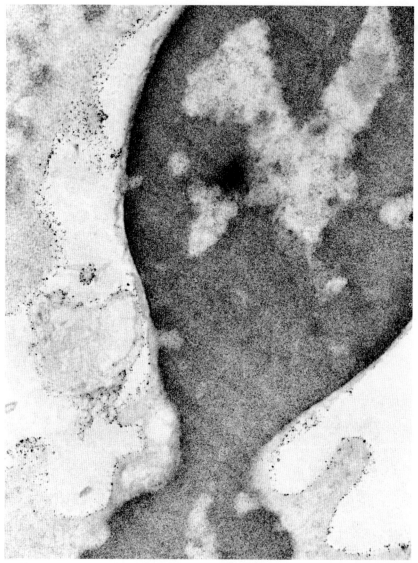

FIG. 13. Bone marrow cells stained with the positive colloid. A late erythroblast is shown representing an early stage in the nuclear expulsion process, in which a part of the nucleus is situated outside the main bulk of cytoplasm, surrounded by a rim of cytoplasm and membrane. The positive colloid on the erythroblast surface is present mainly on that part of the membrane which surrounds the remaining cytoplasm. A part of a leukoid cell (top) shows a membrane heavily coated with dense particles. × 35,000. (Reprinted, with permission, from ref. 26.)

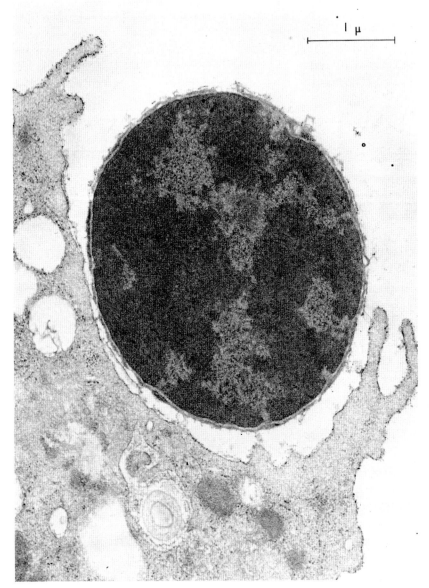

FIG. 14. Thin section of rabbit bone marrow suspension, stained with the positive colloid. Free erythroid nucleus, situated in proximity to a macrophage, is partly surrounded by the cytoplasmic protrusions of the macrophage. The colloid black dots are abundant on the macrophage membrane and scarce on the membrane surrounding the erythroid nucleus. × 23,000. (Reprinted, with permission, from ref. 26.)

The apparent similarity in charge density between rabbit and human red cells, as estimated by the density of colloidal iron particles, is rather surprising in view of the markedly lower electric mobility of rabbit red cells (29). Furthermore, poly-L-lysine, n = 1,000, was required to agglutinate rabbit red cells, while the shorter polymer, n = 100, agglutinated human red cells (11). However, it was estimated that the iron particle covers five to 16 molecules of sialic acid (30), and even if only 10% of these molecules are ionized, there would be sufficient charge to interact with the iron particles (24). Assuming that these charges are uniformly distributed, the equal density of iron particles on the surface of both human and rabbit young red blood cells would be understandable. Apparently, in spite of the difference in their overall surface charge, both contain sufficient charges to saturate the surface with iron particles. The reduction in mobility and in density of labeling with colloidal iron, and the increased rate of agglutination of old cells and of RDE-treated cells can be attributed to the depletion of negative charges in areas on the red cell surface as visualized by the low grain density on the membrane surfaces in the electron micrographs. In this connection, it is interesting to note that if human red cells are labeled with Cr^{51} and injected into rabbits, they disappear from the circulation within the first hr after injection, despite the fact that their surface charge is three times that of rabbit red cells. The recognition of "nonself" seems to overwhelm the charge density signal. On the other hand, Cr^{51}-labeled rabbit red cells treated with neuraminidase and injected into rabbits disappear from the circulation within a day or two (Danon, D. and Marikovsky, Y., unpublished results). The avidity of macrophages in phagocytizing neuraminidase-treated red cells *in vitro* has also been reported (31). It is interesting to note that the neuraminidase-treated cells do not demonstrate any significant changes in other biophysical parameters, such as osmotic fragility, mechanical fragility or density distribution of cells or, at any rate no more than do cells incubated without neuraminidase. These findings, together with the reduced labeling by positively charged colloidal iron particles of the membrane that envelops the nucleus being extruded, lead us to believe that the reduction in surface charge of senescent cells and extruded erythroid nuclei is a major recognition signal that helps the macrophage to distinguish the undesirable "altered or deteriorated self cells."

REFERENCES

1. Marks, P. A. Aspects biochimiques du vieillissement du globule rouge et de l'anemic hemolitique d'origine dedicamenteuse. *Nouv. Rev. franc. Hémat.* **1**: 900, 1960.
2. Bishop, Ch. and Surgenor, D. M. "The red blood cell." New York, Academic Press, 1964.
3. Danon, D. Biophysical aspects of red cell ageing. *Bibl. haemat.* (*Basel*) **29**: 178, 1968.
4. Danon, D. and Marikovsky, Y. Difference de charge electrique de surface entre erythrocytes jeunes et ages. *C.R. Acad. Sci.* (*Paris*) **253**: 1271, 1961.
5. Marikovsky, Y., Danon, D. and Katchalsky, A. Agglutination by polylysine of young and old red blood cells. *Biochim. biophys. Acta* (*Amst.*) **124**: 154, 1966.
6. Yeari, A. Mobility of human red blood cells of different age groups in an electric field. *Blood* **33**: 159, 169.
7. Glaeser, R. M. and Mell, H. C. The electrophoretic behaviour of osmium tetroxide-fixed and potassium permanganate-fixed rat erythrocytes. *Biochim. biophys. Acta* (*Amst.*) **79**: 606, 1964.
8. Haydon, D. A. and Seaman, G. V. E. Electrokinetic studies on the ultrastructure of human erythrocyte. I Electrophoresis at high ionic strengths—the cell as a polyanion. *Arch. Biochem.* **122**: 126, 1967.
9. Sachtleben, P. and Ruhenstroth-Bauer, G. Agglutination and the electrical surface potential of red blood cells. *Nature* (*Lond.*) **192**: 982, 1961.
10. Danon, D., Howe, C. and Lee, L. T. Interaction of polylysine with soluble components of human erythrocyte membranes. *Biochim. biophys. Acta* (*Amst.*) **101**: 201, 1965.
11. Danon, D., Marikovsky, Y. and Kohn, A. Red cell agglutination kinetics: A method of automatic recording with the fragiligraph. *Experientia* (*Basel*) **25**: 104, 1969.
12. Gottschalk, A. and Lind, P. E. Product of interaction between influenza virus enzyme and ovomucin. *Nature* (*Lond.*) **164**: 232, 1949.
13. Cook, G. M., Heard, D. H. and Seaman, G. V. F. Sialic acids and the electrokinetic change of the human erythrocyte. *Nature* (*Lond.*) **191**: 44, 1961.
14. Honig, M. Electrokinetic change in human erythrocytes during adsorption and elution of PR8 influenza virus. *Proc. Soc. exp. Biol.* (*N.Y.*) **68**: 385, 1948.
15. Stone, J.D. and Ada, G. L. Electrophoretic studies of virus-red cell interaction: Relationship between agglutinability and electrophoretic mobility. *Brit. J. exper. Path.* **33**: 428, 1952.
16. Heilmeyer, L. and Bergeman, H. "Blut und Krankheiten." Berlin, Springer Verlag, 1951.
17. Campbell, F. R. Nuclear elimination from the normoblast of fetal guinea pig liver as studied with electron microscopy and serial sectioning techniques. *Anat. Rec.* **160**: 539, 1968.
18. Orlic, D., Gordon, A. S. and Rhodin, J. A. G. An ultrastructural study of erythropoietic induced red cell formation in mouse spleen. *J. Ultrastruct. Res.* **13**: 516, 1965.
19. Pease, D. C. An electron microscopic study of red bone marrow. *Blood* **11**: 501, 1956.
20. Seki, M., Yoneyama, T. and Shirasawa, H. Role of the reticular cells during maturation process of the erythroblast. I. Denucleation of erythroblast by reticular cell: Electron microscopic study. *Acta path. jap.* **15**: 295, 1965.
21. Seki, M., Yoneyama, T. and Shirasawa, H. Role of the reticular cells during maturation process of the erythroblast. II. Further observation on denuculeation process of erythroblast. *Acta path. jap.* **15**: 303, 1965.
22. Skutelsky, E. and Danon, D. An electron microscopical study of nuclear elimination from the late erythroblast. *J. Cell Biol.* **33**: 625, 1967.
23. Zamboni, Z. Electron microscopic studies of blood embryogenesis in humans. II. The hemopoietic activity in the fetal liver. *J. Ultrastruct. Res.* **12**: 525, 1968.

24. GASIC, G. J., BERWICK, I. and SORRENTINO, M. Positive and negative colloidal iron as cell surface electron stains. *Lav. Invest.* **18**: 63, 1968.
25. MARIKOVSKY, Y. and DANON, D. Electron microscope analysis of young and old red blood cells stained with colloidal iron for surface charge evaluation. *J. Cell Biol.* **43**: 1, 1969.
26. SKUTELSKY, E. and DANON, D. Reduction in surface charge as an explanation of the recognition by macrophages of nuclei expelled from normoblasts. *J. Cell Biol.* **43**: 8, 1969.
27. PRANKERD, T. A. J. The ageing of red cells. *J. Physiol. (Lond.)* **143**: 325, 1958.
28. DANON, D. and MARIKOVSKY, Y. Determination of density distribution of red blood cells. *J. Lab. clin. Med.* **64**: 668, 1964.
29. EYLAR, E. H., MADOFF, M. A., BRODY, O. V. and ONCLEY, J. L. The contribution of sialic acid to the surface charge of the erythrocyte. *J. biol. Chem.* **237**: 1992, 1962.
30. KRAEMER, P. M. Sialic acid of mammalian cell lines. *J. Cell. Physiol.* **67**: 23, 1966.
31. GARDNER, E., WRIGHT, C. S. and WILLIAMS, B. S. The survival of virus-treated erythrocytes in normal and splenectomized rabbits. *J. Lab. clin. Med.* **58**: 743, 1961.

RED CELL METABOLISM IN HEAT-ACCLIMATED GOLDEN HAMSTERS

NAOMI MEYERSTEIN and YAIR CASSUTO

Department of Environmental Physiology,
Negev Institute for Arid Zone Research, Beersheba, Israel

Different aspects of the effect of heat on red blood cells have been studied. *In vitro* heating of red cells causes increased osmotic fragility (1) and morphological changes, such as the appearance of spherocytes and the fragmentation of red blood cells (2). Osmotic fragility of isolated red cells incubated at 0 to 40 C decreases with the temperature rise (3). Karle has demonstrated that experimental pyrexia reduces red cell survival time (4). *In vivo* pyrexia, induced by injecting pyrogens, or by external heating, increases osmotic fragility of red cells in rabbit and in man (5).

The *in vivo* studies were performed on hyperthermic mammals. It was thus decided to study the changes in red blood cells of heat-acclimated animals. Hamsters become heat-acclimated after being kept in the climatic chamber for three weeks. These hamsters have lower metabolic rates than controls (6), and this improves their tolerance to acute heat exposure (7). It has also been demonstrated that acclimation decreases both the respiratory rates of isolated liver mitochondria and the catabolic reactions of its carbohydrate metabolism. Therefore, it was of interest, too, to study the glucose metabolism of the red cell as an example of another type of tissue which depends only on glucose metabolism.

This communication reports preliminary results of these investigations.

MATERIALS AND METHODS

Adult male hamsters (110 ± 10 g) were randomly divided into groups; one group (control) was maintained at room temperature (20 to 23 C), the second group (heat-acclimated) was kept in the climatic chamber at 34 to 36 C with relative humidity at 25 to 40%. The acclimation period in the climatic chamber was 21 to 30 days. A third group (deacclimated) was kept in the climatic chamber for 21 days, returned to the control room, and sacrificed within eight to 11 days after removal from the climatic chamber. The animals were killed by decapitation and the blood was collected into heparinized tubes. Peripheral smears were taken. Freshly drawn blood was collected directly into cold perchloric acid and assayed for adenosine triphosphate (ATP) level (8). Red blood cells were isolated (9) and then incubated in glycyl-glycine buffer (pH 7.8) at 37 C. Glucose was determined by the glucose oxidase method (10), and lactate with lactic dehydrogenase and nicotinamide adenine dinucleotide (11). Osmotic fragility studies were performed according to Suess et al. (12), except that buffered electrolyte solutions were used (13). Hemolysis percentage was calculated by comparison with hemolysis in distilled water.

RESULTS

The blood smears of both control and heat-acclimated groups showed similar red cell morphology. No appreciable spherocytosis was noted in the peripheral blood smears of heat-acclimated hamsters.

Body temperatures of the control animals (37.9 ± 0.2 C) were similar to those of the heat-acclimated animals (38.0 ± 0.1 C).

Preliminary experiments established linear progression rates of glucose utilization and lactate production between the 15th and the 90th min of incubation. As an additional control, parallel studies were made on human red cells. Table 1 demonstrates a significant ($P < 0.01$) increase of 15% in

TABLE 1. *Glucose uptake, lactate production and ATP levels in heat-acclimated and control hamsters* [a]

	Control	Heat-acclimated	Percentage difference	Significance value
Glucose uptake (μmole/g Hb per hr) (20)	14.27 ± 0.4	16.41 ± 0.7	15%	$P < 0.01$
Lactate production (μmole/g Hb per hr) (20)	32.25 ± 1.62	35.50 ± 1.43	10%	$P < 0.1$
Lactate/glucose (20)	2.25 ± 0.06	2.18 ± 0.07	3%	n.s.
ATP (μmole/10^{10} RBC) (8)	1.07 ± 0.02	0.86 ± 0.01	20%	$P < 0.001$

Number of animals in each group appears in parenthesis.
[a] All values are means $\pm$ SE.

glucose uptake in red cells of the heat-acclimated over that of the control hamsters. Lactate production increased by 10%, but the increase was not statistically significant ($P < 0.1$). The ratio of lactate/glucose did not change.

Fig. 1 shows the marked increase in osmotic fragility in red cells of heat-acclimated hamsters. The osmotic fragility curve of the red cells of deaccli-

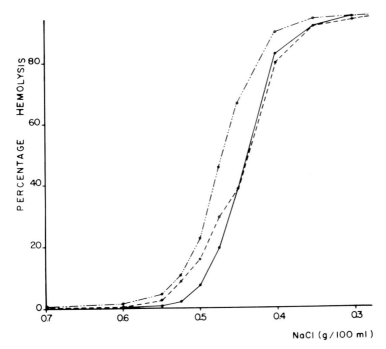

FIG. 1. Red cell osmotic fragility of heat-acclimated hamsters (— · · —), deacclimated (– – – –), and control hamsters (—). Mean results of 33 control, 33 heat-acclimated and eight deacclimated hamsters.
Abscissa: tonicity expressed as NaCl in g/100 ml.
Ordinate: hemolysis as percentage of hemolysis in water.

mated hamsters overlaps that of the controls at concentrations up to 0.45% NaCl, while at higher concentrations it is located between the curves of the heat-acclimated and control animals.

When the same data are plotted as hemolytic increments (Fig. 2), it becomes apparent that the control curve has a sharply peaked monophasic form, whereas the heat-acclimated curve starts at higher osmotic concentrations, and has a flattened peak. The deacclimation curve shows one peak

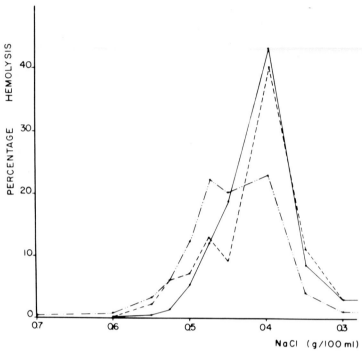

FIG. 2. Red cell osmotic fragility of heat-acclimated hamsters (— — — —), deacclimated (– – – –), and control hamsters (—). Mean results of 33 control, 33 heat-acclimated and eight deacclimated hamsters.
Abscissa: tonicity expressed as NaCl in g/100 ml.
Ordinate: hemolysis as percentage of hemolysis in water.
Same results as in Fig. 1 expressed as hemolytic increments.

at low concentrations (identical to that of the control), and another which is more similar to the heat-acclimated curve. The increased osmotic fragility is probably due to metabolic changes occurring in the heat-acclimated animal. Table 1 demonstrates clearly the 20% ($P < 0.001$) reduction in red cell ATP content observed in freshly drawn blood of heat-acclimated hamsters.

DISCUSSION

It has been shown that both *in vitro* heating of red cells and *in vivo* pyrexia cause increased osmotic fragility. As shown by the data presented here, osmotic fragility increases also by heat acclimation. This increase is highly significant ($P < 0.001$) at the concentration range of 0.85 to 0.4% NaCl, but has no significance at concentrations lower than 0.4% NaCl. These animals are normothermic, in contrast to the hyperthermic rabbit and man in the experimental pyrexia studies, so that their increased fragility cannot be due to hyperthermia. As Fig. 2 demonstrates, the red cells of the control group belong to a distinct, uniform population, whereas the cells of the heat-acclimated animals have a broader response to hypotonicity, which suggests the existence of different populations. This effect becomes more pronounced in the deacclimated animals, where at least two populations exist. Hence it can be hypothesized that during the deacclimation period new cells, with control group characteristics, are produced. Another possible explanation is that during the deacclimation period certain plasmatic factors augmenting osmotic fragility are removed. The increased osmotic fragility may correlate with metabolic changes in the red cell. The glucose uptake of the red cells as opposed to that of the liver, in heat-acclimated hamsters, did not decrease but rose significantly. This increase could be due to an increase in reticulocyte count (14), although such reticulocytosis would have caused decreased fragility (15). As reticulocyte counts in the two groups were similar, this does not explain the higher glucose consumption. Red cell survival in the hamster is 60 to 78.5 days (16); thus it may be argued that the red cell population in the experimental group at the time of killing included red cells already in existence on introduction into the climatic chamber. Such cells could have been damaged at that initial heat exposure when the animals were not yet acclimated. Fragility studies and reticulocyte counts were therefore performed on the red cells of hamsters kept for six months in the chamber. The results were identical to those found in heat-acclimated hamsters which had been kept in the climatic chamber for only 21 days. The entire red cell population at the time of sacrifice had been produced solely during the six-month acclimation.

To summarize, it has been shown that the red cells of heat-acclimated hamsters have low ATP levels, increased glucose uptake and increased osmotic fragility. Thus, these red cells manifest changed patterns of metabolism and fragility. Some characteristics of these cells resemble those of red cells in hereditary spherocytosis and pyruvate kinase deficiency (17, 18). The primary cause of the changes in the red cells of heat-acclimated hamsters is not known and is being investigated.

The authors are grateful to Prof. D. Danon for helpful discussion. We also wish to thank Mrs. Cynthia Bellon for her assistance in preparing the manuscript.

Supported in part by Grant BDPEC–OH–ISR–7 from the U.S. Department of Health, Education and Welfare.

REFERENCES

1. ISAACS, R., BROCK, B. and MINOT, G. R.   The resistance of immature erythrocytes to heat. *J. clin. Invest.* **1**: 425, 1925.
2. SCHULTZE, M.   Ein heizbarer Objecttisch und seine Verwendung bei Untersuchung des Blutes. *Arch. mikr. Anat.* **1**: 1, 1865.
3. PARPART, A. K., LORENZ, P. B., PARPART, E.R., GREGG, J. R. and CHASE, A.M.   The osmotic resistance (fragility) of human red cells. *J. clin. Invest.* **26**: 636, 1947.
4. KARLE, H.   Elevated body temperature and the survival of red blood cells. *Acta med. scand.* **183**: 587, 1968.
5. KARLE, H.   Effect on red cells of a small rise in temperature. *In vitro* studies. *Brit. J. Haemat.* **16**: 409, 1969.
6. CASSUTO, Y.   Metabolic adaptations to chronic heat exposure in the golden hamster. *Amer. J. Physiol.* **214**: 1147, 1968.
7. CASSUTO, Y. and CHAFFEE, R. R. J.   Effect of prolonged heat exposure on the cellular metabolism of the hamster. *Amer. J. Physiol.* **210**: 423, 1966.
8. ADAM, H.   Adenosine-5'-triphosphate determination with phosphoglycerate kinase, in: Bergmeyer, H.U. (Ed.), "Methods of enzymatic analysis," revised edn. New York, Academic Press, 1965, p. 539.
9. BUSCH, D. und PELZ, K.   Erythrozytenisolierung aus Blut mit Baumwolle. *Klin. Wschr.* **44**: 983, 1966.
10. BERGMEYER, H. U. and BERNT, E.   Determination of D-glucose with glucose oxidase and peroxidase, in: Bergmeyer, H. U. (Ed.), "Methods of enzymatic analysis," revised edn. New York, Academic Press, 1965, p. 123.
11. HOHORST, J. H.   Determination of L(+) lactate with lactic dehydrogenase and DPN, in: Bergmeyer, H. U. (Ed.), "Methods of enzymatic analysis," revised edn. New York, Academic Press, 1965, p. 266.
12. SUESS, J., LIMENTANI, D., DAMESHEK, W. and DOLLOFF, M.   A quantitative method for the determination and charting of the erythrocyte hypotonic fragility. *Blood* **3**: 1290, 1948.
13. WINTROBE, M. M.   "Clinical hematology," 6th edn. Philadelphia, Lea and Febiger, 1967.
14. BERNSTEIN, R. E.   Alterations in metabolic energetics and cation transport during ageing of red cells. *J. clin. Invest.* **38**: 1572, 1959.
15. MARKS, P. A. and JOHNSON, A. B.   Relationship between the age of human erythrocytes and their osmotic resistance: A basis for separating young and old erythrocytes. *J. clin. Invest.* **31**: 1542, 1958.
16. BROCK, M. A.   Production and life span of erythrocytes, during hibernation in the golden hamster. *Amer. J. Physiol.* **198**: 1181, 1960.
17. JANDL, J. H.   Hereditary spherocytosis, in: Beutler, E. (Ed.), "Hereditary disorders of erythrocyte metabolism." New York, Grune and Stratton Inc., 1967, p. 209.
18. TANAKA, K. R. and VALENTINE, W. N.   Pyruvate kinase deficiency, in: Beutler, E. (Ed.), "Hereditary disorders of erythrocyte metabolsim." New York, Grune and Stratton Inc., 1967, p. 229.

HEINZ BODY FORMATION IN RED CELLS OF THE NEWBORN INFANT

E. KLEIHAUER,[*] A. BERNAU, K. BETKE[*] and M. KELLER

Departments of Pediatrics[*] and Hematology, Universities of Ulm and Munich, West Germany

Early observations by Willi (1) and Gasser (2) have shown that newborn infants, especially prematures, are more liable to form Heinz bodies within their erythrocytes than are adults. The enhanced Heinz body formation becomes evident following administration of compounds capable of acting as oxidizing agents causing hemolytic Heinz body anemia (3–6).

The increased degradation of Hb in RBC of newborn subjects under experimental *in vitro* conditions was first described by Künzer et al. (7) and confirmed by others (8–14). The mechanisms leading to this phenomenon cannot be fully explained either by the decreased activities of some red cell enzymes (15, 16) or by the presence of fetal Hb (17).

The present study was undertaken a) to investigate some fundamental conditions governing Heinz body formation in erythrocytes from adult and cord blood, and b) to answer the question of whether adult RBC would change their susceptibility to form Heinz bodies in the circulation of newborns after exchange transfusion. The latter problem was initiated by the observation of Irle (18) who reported three cases of acute toxic hemolytic Heinz body anemia in newborn infants; one of them had a history of exchange transfusion three weeks previously.

MATERIALS AND METHODS

Freshly drawn venous blood samples were incubated with β-acetylphenylhydrazine according to the procedure of Beutler et al. (19). Heinz bodies were stained with Nile blue sulfate. The number of RBC containing Heinz bodies was counted among 1,000 erythrocytes. The number of Heinz bodies/single red cell was not taken into account.

Data on Heinz body formation in RBC of Rh-hemolytic infants treated with exchange transfusion are based on 11 observations. Blood samples were collected in acid citrate dextrose (ACD) solution before and immediately after blood exchange. Studies were continued during subsequent hours and days. Donor blood in ACD solution served as a control. Samples were drawn from the bag before and at the end of exchange transfusion. A second adult blood sample was included as an additional control.

RESULTS

Under normal atmospheric conditions, Heinz body formation in the presence of β-acetylphenylhydrazine follows a characteristic course (Fig. 1). After a latent period, the number of RBC containing Heinz bodies increases rapidly. The time required to produce Heinz bodies in 50% of the red cells at a final concentration of 10.5 mmole β-acetylphenylhydrazine is approximately 1½ hr for cord blood and 4 hr for adult blood. The time lapse between onset of Heinz body formation and their appearance in all the RBC containing Heinz bodies was equal in both samples.

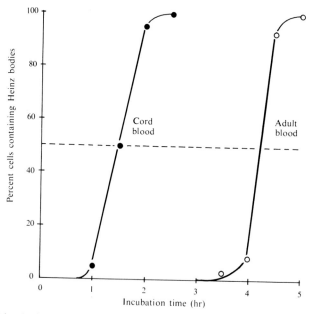

FIG. 1. Heinz body formation in RBC of cord blood and adult blood during incubation with β-acetylphenylhydrazine. Final concentration 10.5 mM. Note the short latent period in cord blood (approx. 1 hr) as compared with adult blood (approx. 3½ hr).

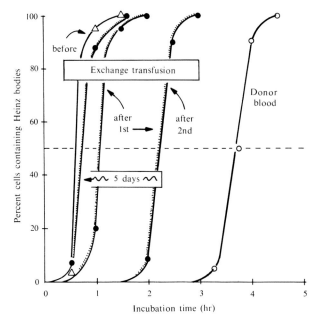

FIG. 2. Heinz body formation in RBC before and after exchange transfusion. Note the differences between donor blood and transfused RBC five days after the second exchange transfusion, as well as the differences between the first and second blood exchange.

Investigations by Waller and Löhr (20) revealed that Heinz body formation is highly dependent on pH. Oxidative denaturation of Hb is accelerated by increasing the pH with an optimum at pH 8.0. In our own experiments the pH at the beginning was always low in the cord blood samples as compared with adult blood under the same conditions. Only at the end of the incubation period were the pH values almost equal in both blood samples.

Experiments to clarify the influence of different atmospheric conditions failed to produce Heinz bodies in the absence of oxygen (carbon monoxide, nitrogen) in cord blood as well as in adult blood. Increase in oxygen pressure was correlated with an increased Heinz body formation. The differences between fetal and adult RBC again consisted of differences in the time necessary for the onset of Heinz body formation.

Studies concerning Heinz body formation in RBC of infants before and after exchange transfusion gave quite unexpected results. Without exception, the transfused adult donor cells in the circulation of newborn infants showed a higher susceptibility to Heinz body formation than the same RBC before and after storage at room temperature during exchange transfusion.

The increase in Heinz body formation could be demonstrated immediately after the blood exchange was terminated. During subsequent hours and days, Heinz body formation was approximately as fast as in the newborns' own RBC. The same phenomenon was demonstrable in the course of two exchange transfusions (Fig. 2) in which fetal RBC were almost completely replaced by adult RBC.

DISCUSSION

Experimental data from the present study indicate that the Heinz body production in fetal RBC requires a shorter incubation time than in adult cells. Furthermore, it is evident from the data presented that the enhanced Heinz body formation in RBC of newborn infants cannot be explained as an effect of pH under *in vitro* conditions.

It is well known that RBC of newborn infants have a number of different properties as compared with adult RBC (13, 15, 21). This is not only true for the Hb pattern and enzyme activities but also for the membrane composition. Several mechanisms have been postulated to be responsible for the increased susceptibility of fetal RBC to form Heinz bodies. Glutathione instability (10, 22) as well as catalase deficiency (23, 24) can be ruled out as factors causing this phenomenon. Recent studies confirmed the role of glutathione peroxidase as the enzyme affording protection against oxidative denaturation (25, 26), although in RBC of newborn infants decreased activities of this enzyme have been observed (16, 27); the significance, however, of these findings is difficult to assess. According to data given by Gross et al. (16), approximately 40% of newborn infants have activities of glutathione peroxidase in RBC within the normal adult range, while Necheles et al. (27) repeatedly found decreased activities in newborns. Our results concerning Heinz body formation in adult RBC before and after exchange transfusion seem to argue against a direct glutathione peroxidase in determining the extent of the phenomenon. There is no evidence which demonstrates that the enzyme activity decreases in circulating RBC while it remains unaffected in the same RBC stored at room temperature during the time of exchange transfusion. This awaits further investigation.

Fetal Hb content of RBC seems to be of minor significance for the enhanced Heinz body formation. Experimental data of Ulukutlu et al. (17) indicate a slightly increased Heinz body production in fetal cells containing predominantly fetal Hb as compared to fetal cells containing predominantly adult Hb. However, the latter are much more susceptible than RBC from adults containing adult Hb. The pronounced differences between both types of cells containing adult Hb disappear after exchange transfusion.

An additional point to be considered in this connection is the role of the RBC membrane. There are differences in the phospholipid composition between RBC of adults and newborn infants. However, quantitative data differ widely in different studies (28–30). Recent studies by Hürter et al. (to be published) indicate that the total content of RBC and plasma phospholipids is significantly decreased when compared with corresponding data in adults. Thus the differences could be explained by an exchange of phospholipids between plasma and erythrocyte membrane (31, 32). Correlations between phospholipid and thiol degradation and oxidative damage of RBC, as well as the protective effect of vitamin E have been reported by Jacob and Lux (33).

As a working hypothesis, we propose that the change in susceptibility of adult RBC to Heinz body formation may be due to alterations in the phospholipid content which occur during circulation of the donor blood in the newborn. Thus adult RBC may acquire properties of fetal cells, at least as far as membrane characteristics are concerned. Studies are in progress to test this hypothesis.

Supported by the Deutsche Forschungsgemeinschaft.

REFERENCES

1. WILLI, H.   Innenkörperbildung durch Elkosin und spontane Innenkörperbildung. *Schweiz. med. Wschr.* **77**: 243, 1947.
2. GASSER, C.   Die hämolytische Frühgeburtenanämie mit spontaner Innenkörperbildung: ein neues Syndrom, beobachtet in 14 Fällen. *Helv. paediat. Acta* **8**: 491, 1953.
3. BETKE, K.   Erworbene hämolytische Erkrankungen des Neugeborenen. in: *VII Freiburg Symposium on Hämolyse und hämolytische Erkrankungen, Oktober,* 1959, Berlin, Springer, 1961.
4. BETKE, K.   Toxische hämolytische Anämien, in: Opitz, H. and Schmid, F. (Eds.), "Handbuch der Kinderheilkunde," 6th edn. Berlin, Springer, 1967, p. 945.
5. DAWSON, J. P., THAYER, W. W. and DESFORGES, J. F.   Acute hemolytic anemia in the newborn infant due to naphthalene poisoning: Report of two cases, with investigations into the mechanism of the disease. *Blood* **13**: 1,113, 1958.
6. VALAES, T., DOXIADIS, S. A. and FESSAS, P.   Acute hemolysis due to naphthalene inhalation. *J. Pediat.* **63**: 904, 1963.
7. KÜNZER, W., AMBS, E. and SCHNEIDER, D.   Untersuchungen zur Heinzkörperbildung in Neugeborenenerythrozyten. *Z. Kinderheilk.* **74**: 652, 1954.
8. SANSON, G., BORRONE, C. and ROVEI, S.   Suscettibilità degli eritrociti a formare corpi di Heinz in vitro in condizioni normali (neonati e lattanti) e patologiche (favismo e talassemia). *Boll. Soc. ital. Biol. sper.* **34**: 1561, 1958.
9. VEST, M.   Der Einfluz von Naphthohydrochinoderivaten (wasserlöslichen Vitamin-K-Ersatzpräparaten, Synkavit) auf Erythrozytenabbau und Regeneration bei

Frühgeburten und auf das Glukuronidbildungsvermögen der Leber in vitro. *Schweiz. med. Wschr.* **88**: 969, 1958.
10. TJOA, G. T. Untersuchungen über Heinzkörperbildung und Glutathionstabilität in Erythrozyten von Neugeborenen. *Z. Kinderheilk.* **84**: 484, 1960.
11. BROWN, A. K. and BUQUIR, S. L. Spontaneous Heinz body formation in full-term and premature infants. *Amer. J. Dis. Child.* **102**: 589, 1961.
12. GROSS, R. T. and SCHROEDER, E. A. R. The relationship of triphosphopyridine nucleotide content to abnormalities in the erythrocytes of premature infants. *J. Pediat.* **63**: 823, 1963.
13. KLEIHAUER, E. Fetales Hämoglobin und fetale Erythrozyten. *Arch. Kinderheilk.* **53**: 1966.
14. BORGES, A and DESFORGES, J. F. Studies of Heinz body formation. *Acta haemat. (Basel)* **37**: 1, 1967.
15. BROWN, A. K. Erythrocyte metabolism and hemolysis in the newborn. *Pediat. Clin. N. Amer.* **13**: 879, 1966.
16. GROSS, R. T., BRACCI, R., RUDOLPH, N., SCHROEDER, E. and KOCHEN, J. A. Hydrogen peroxide toxicity and detoxification in the erythrocytes of newborn infants. *Blood* **29**: 481, 1967.
17. ULUKUTLU, L., KLEIHAUER, E. and BETKE, K. The behaviour of foetal haemoglobin in the enhanced Heinz body formation in red cells of newborn infants. *Acta paediat. scand.* **55**: 473, 1966.
18. IRLE, U. Akute hämolytische Anämie durch Naphthalin-Inhalation bei zwei Frühgeborenen und einem Neugeborenen. *Dtsch. med. Wschr.* **89**: 1798, 1964.
19. BEUTLER, E., DERN, R. and ALVING, A. The hemolytic effect of primaquine. VI. An in vitro test for sensitivity of erythrocytes to primaquine. *J. Lab. clin. Med.* **45**: 40, 1955.
20. WALLER, H. D. and LOHR, G. W. Neue Ergebnisse zum Mechanismus der Heinzkörperbildung in Erythrozyten. *Folia haemat. (Frankfürt)* **8**: 1, 1963.
21. OSKI, F. A. and NAIMAN, J. L. Hematologic problems in the newborn, in: Schaffer, A. J., (Ed.), "Major problems in clinical pediatrics." Philadelphia, W. B. Saunders, Co., 1966, IV.
22. ZINKHAM, W. H. An in vitro abnormality of glutathione metabolism in erythrocytes from normal newborns: Mechanism and clinical significance. *Pediatrics* **23**: 18, 1959.
23. KUNZER, W. Zur Frage der Heinzkörperbildung in Neugeborenenerythrozyten. *Folia haemat. (Lpz.)* **73**: 405, 1956.
24. BRENNER, S. and ALLISON, A. C. Catalase inhibition: a possible mechanism for the production of Heinz-bodies in erythrocytes. *Experientia (Basel)* **9**: 381, 1953.
25. MILLS, G. C The purification and properties of glutathione peroxidase of erythrocytes. *J. biol. Chem.* **234**: 502, 1959.
26. COHEN, G. and HOCHSTEIN, P. Glucose-6-phosphate dehydrogenase and detoxification of hydrogen peroxide in human erythrocytes. *Science* **134**: 1756, 1961.
27. NECHELES, T. F., BOLES, T. and ALLEN, D. M. Erythrocyte glutathione peroxidase deficiency and hemolytic disease of the newborn infant. *J. Pediat.* **72**: 319, 1968.
28. CROWLEY, J., WAYS, P. and JONES, J. W. Human fetal erythrocyte and plasma lipids. *J. clin. Invest.* **44** ; 989, 1965.
29. HURTER, P. and SCHROTER, W. Erythrocyte and plasma phospholipids: Studies in adults, neonates and in patients with Rh erythroblastosis and hemolytic anaemias. *Europ. Soc. Pediat. Res., Interlaken, Sept.* 1969.
30. NEERHOUT, R. C. Erythrocyte lipids in the neonate. *Pediat. Res.* **2**: 172, 1968.
31. TARLOW, A. R. and MULDER, E. Phospholipid metabolism in rat erythrocytes: quantitative studies of lecithin biosynthesis. *Blood* **30**: 853, 1967.
32. REED, C. F. Phospholipid exchange between plasma and erythrocytes in man and the dog. *J. clin. Invest.* **47**: 749, 1968.
33. JACOB, H. S. and LUX, S. E. Degradation of membrane phospholipids and thiois in peroxide hemolysis: studies in Vitamin E deficiency. *Blood* **32**: 549, 1968.

DISCUSSION

R. F. Rieder (*USA*): The papers presented by Drs. E. M. and N. Kosower are open for discussion.

J. Mager (*Israel*): What does the analogue of δ-aminolevulinic acid (ALA), tested by you in experimental porphyria, do to a normal animal? How does it affect the susceptibility of a normal animal to ALA? Is this in fact responsible for the neurological signs of porphyria?

N. Kosower (*USA*): There was no effect of the analogue on normal animals, except for a slight change in the amount of porphobilinogen (PBG) excreted. As to the second question, we have not studied this. We would like to have a model where one would be able to give ALA to a normal animal and see what happens. However, technically it's not easy. People have claimed in the past that ALA itself is not toxic, apart from some transient photosensitivity. In acute intermittent porphyria, however, photosensitivity does not occur. Since ALA injected in large amounts into normal animals is removed very quickly by conversion to PBG and the rest is excreted, it is hard to keep a continuous high level of ALA in the blood.

E. M. Kosower (*USA*): I would only like to add that we did measure the rates of reaction of ALA with pyridoxal. It rapidly forms a Schiff base, just as do all other amino acids. In fact, the rates are very similar to those of other amino acids. We also looked at the rate at which this Schiff base disappears and hydrolyzes. Under neutral (physiological) conditions its half-life is in the order of 12 sec, which means that only a small fraction of the material would be transported from the liver, where we believe it is formed, to the brain. In summary, the proper model for determining what ALA added from outside does to an organism is to provide an increased steady state concentration of ALA, because we think that it is the increased level of the pyridoxal-ALA combination that leads eventually to the consequences, and we cannot simulate this by one single dose.

J. Mager: I understood that the diazo inhibits the synthetic reaction or the dehydrase reaction. It seems also that the formation or the utilization of ALA is inhibited. If so, there should be an accumulation of ALA.

E. M. Kosower: I pointed out that there was some inhibition of the synthetic reaction. This is a possibility of course. What I suggested is that the analogue be used to increase the steady state concentration of ALA and thus lead eventually to a porphyric animal.

R. F. RIEDER: I asked the question whether or not this compound inhibited heme-synthesis and whether it would be useful in treating diseases like erythropoietic protoporphyria, in which there is an excess of heme synthesis.

E. M. KOSOWER: This is certainly a good point. I hope everyone remembers that this is a newly synthesized compound; it was made once before but not studied biologically. It cannot be used in man and we do not have animals with such diseases.

R. F. RIEDER: There is now some *in vitro* evidence that ALA is toxic to the neuromuscular function. Drs. D. S. Feldman and R. D. Levere at Downstate Medical Center have studied the *in vitro* effects of ALA and have shown that this affects acetylcholine release in nerve muscle preparations.

D. DANON (*Israel*): Would you elaborate on the kind of damage that you observed in red cells other than hemolysis?

N. KOSOWER: This refers to the situation where additional challenge in the form of free radicals is generated near the membrane. After GSH oxidation, prior to frank hemolysis *in vitro*, we found an increased osmotic fragility and a decrease in cell density. This would indicate an intake of water into the cell. Preliminary studies show that the ATP content and adenosine triphosphatase activity of the cells was unchanged. We do not know the exact molecular site of the effect, but from the biophysical parameters measured, very little intracellular damage was observed. The hemoglobin inside the cell is fully functional, as measured by the oxygen dissociation curve. The ghosts of these cells show tears and holes under the electron microscope. (A large part of this work was done in Prof. Danon's laboratory during our sabbatical year at the Weizmann Institute.) I mentioned before that we have used the methods developed by Danon to measure the surface charge of the red cells. If we take an oxygenated red cell where most of the damage is intracellular, we see a very mild effect on the membrane that consists only of a mild reduction in surface charge as evidenced by a reduction in electric mobility and an enhanced rate of agglutination by polylysine. Loss of charge might be of importance in removal of mildly damaged cells by the spleen; however, in the second case the cells are probably injured to such a degree that they can burst inside the circulation.

E. GOLDSCHMIDT (*Israel*): Drs. E. M. Kosower and D. Danon agree about the membrane alterations. Would you tell us about it? One of you gets it by hydration and the other by loss of sialic acid. So I guess that you have made sure that Dr. N. Kosower's results are not due to loss of sialic acid.

D. DANON: In the intact red blood cell, the main surface charge is due to N-acetyl-neuraminic acid, the carboxylic group of which was estimated to be approximately 60 to 80 Å of what we see as membrane in the electron microscope. Any deterioration of the type demonstrated by N. Kosower may cause structural changes in the membrane that would probably alter the position of the carrier of the negative charge in the membrane. If the change is no longer oriented to the surface, it may be ineffective. The experimental fact is that we see a change in electric mobility and agglutinability, and in several experiments less negative charge labeling with colloidal iron was noted. This is not necessarily due to destruction of the carboxylic groups. They may be hidden as a result of deterioration of the membrane. This deterioration can be visualized by electron microscopy.

E. KLEIHAUER (*West Germany*): Is it necessary to wash out the azoester after incubation with red cells, or is it unstable? The next question concerns the dose effect. You need only small amounts of azoester to lower GSH activity, but you need an excess of it to produce hemolysis and Heinz bodies. Is the hemolysis due only to a very lowered GSH content, or is there also an effect on the membrane?

N. KOSOWER: In order for the red cell to be damaged, one would need two factors: firstly a lowering of the GSH to a subcritical level and then an additional challenge to the red cell at a time when the cell is depleted or almost depleted of GSH. By this I mean that if you take small amounts of a drug like azoester, you do not get any demonstrable damage. If you add more than necessary to oxidize the GSH, free radicals are produced that damage the cell. On the other hand, if you take a reagent which just oxidizes the GSH and by itself cannot be used as an additional challenge, nothing seems to happen, as with the diamide already mentioned. This mechanism does not apply to other biological systems.

J. C. KAPLAN (*France*): Is glutathione the only possible target for azoester?

N. KOSOWER: In the red blood cell, the target for azoester and the other studied compounds appears to be glutathione. It seems that thiol-containing small molecules are very reactive. They also include cysteine and possibly ergothionine. We designate them all GSH because we measure it by the 5'5' dithiobis 2 nitrobenzoac acid method. However, the amounts of the other two compounds as compared to GSH are negligible. The azoester does not react with the SH groups of hemoglobin. This we have done. If we take deoxyhemoglobin, excluding oxygen and thus free radicals from the system, the azoester will not react with hemoglobin. I would say that most protein SH groups under the biological conditions used would probably be very unreactive towards these compounds.

E. M. KOSOWER: I said something about how the azoester was discovered. I think it would take a very wise man to have guessed that such a molecule would exist. We came across it following the clue given by Beutler in his work on acetylphenylhydrazine. It happens that the mechanism of the reaction of azoester with glutathione involves two steps; one is an addition to the molecule and the second is a reaction to form a disulfide. These steps have actually been demonstrated chemically in a number of ways. It turns out that the first step is best for small molecules and for acidic thiol groups, glutathione and cysteine. Large molecule thiols, like protein thiols, do not seem to react as rapidly. The second step, where you form the disulfide, is really what carries the oxidation to completion. This step is very difficult for large molecules because of steric hindrance. Therefore, by some rather unique set of chemical circumstances, we have a molecule which prefers to oxidize glutathione and other small thiols, but not large thiols.

Dr. N. Kosower has just mentioned the fact that the azoester does not react with hemoglobin. We have more recently done an experiment with papain, an SH enzyme even more acidic than glutathione. Therefore, it ought to be more reactive than glutathione and we have shown that it is so unreactive that we cannot even demonstrate it. These observations and the fact that the diamide, which only reacts with SH groups, reacts in almost perfect stoicheiometry with GSH in the red blood cell, makes it very hard to believe that many other thiol groups are involved in the reaction. The effect of the diamide was the same in all the biological systems we have seen.

R. F. RIEDER: I have done a few studies on the formation of Heinz bodies in red cells with unstable hemoglobins, using some of the redox dyes, and in normal cells, using acetylphenylhydrazine. By incubating the cells in the presence of cyanide, one is able to prevent the formation of Heinz bodies by redox dyes, such as brilliant cresyl blue and new methylene blue. I have attributed it to the stabilization effect of cyanide on methemoglobin. It has quite a striking effect when tested by the heat-denaturation method. The cyanide-treated cells show less hemolysis and little morphologic alteration; and I was wondering how you would explain this in terms of the membrane when, as far as I know, the cyanide would have no effect on the membrane.

E. M. KOSOWER: Cyanide protects cells from the formation of Heinz bodies. Let us ask ourselves what Heinz body formation is. We don't know precisely, but I think we can say in very general terms that some reactive compounds coming from outside the cell, or perhaps present within the cell, interact with oxygen or oxyhemoglobin within the cell. It is oxidized to some reactive species, but we imagine that there could be many reactive species. In our work, we deal with one kind of reactive species, a phenyl radical, but there is no reason to limit it to that. In the case of dyes you may transfer an electron either to the dye, which then may react with oxygen, or take it away from the dye giving free radicals of a different type from the ones we have, and these then react with hemoglobin to cause some chemical change. Abstraction of a hydrogen atom is the most likely one. This reactive hemoglobin can do many things, one of which is to react with oxygen. We do not know what chemical reactions take place leading to hemoglobin instability. If, however, you begin with an unstable hemoglobin then you can go more easily to a denatured hemoglobin and finally to the kind of intermolecular interactions, like those in sickle cell anemia, which lead to their separation as gross microscopically visible bodies. In short there is a general process — reactive molecules can react with either oxyhemoglobin or other reagents to generate free radicals or other reactive molecules. These react with hemoglobin to give unstable hemoglobins, the latter denature and then they coalesce to form Heinz bodies. I think that Dr. Kleihauer's work has shown almost unequivocally that there is something in the circulation of newborns which is more sensitive to oxidation. One should try to find this compound. We have done some work that suggests that

DOPA can cause a similar change in GSH. I would like to make an additional comment. To attack the membrane we have to deprive the cell of oxygen. We do it in the red blood cell by putting in carbon monoxide which has a much higher affinity for hemoglobin than oxygen. In our system oxygen is required to produce free radicals. Since oxygen enters the cell from the outside, there is a concentration gradient to the outside. Therefore, free radicals are generated largely near the membrane of the cell. The damage produced can be seen under the electron microscope and hemolysis ensues. If the same thing is done without depriving the cell of oxygen, the free radicals are generated inside the cell and we get mainly intracellular damage. By regulating where we generate the reactive species, which may or may not be a free radical, we can either attack hemoglobin inside or attack the membrane. As for your opinion that cyanide stabilizes methemoglobin, I think that is very reasonable.

B. RAMOT (*Israel*): Dr. Meyerstein's paper is now open for discussion. The changes of surface to volume ratio or spherocytosis would not account for the changes of the red cell as demonstrated in osmotic fragility. Apparently the damage should be somewhere in the membrane. I would ask two questions: Is there any difference in protein synthesis of these cells? And, is there any difference in the life span of these cells?

N. MEYERSTEIN (*Israel*): We have not done protein synthesis studies or survival time determinations. In studies carried out by Karle in pyrexic animals, red cell survival time was found to be reduced. We also consider the primary defect to be in the membrane and, in order to elucidate whether there is a plasmatic factor in the heat-acclimated hamsters which affects the red cell membrane, we performed cross incubations. The plasma of heat-acclimated hamsters had no effect on normal red cells nor did we find any improvement on incubating red cells of heat-acclimated animals with control plasma. I would like to add that these experiments were not conclusive, as the incubations were only short.

E. M. KOSOWER: Is the temperature in the extremities of the animal the same as in the chamber?

N. MEYERSTEIN: We did not measure the skin temperature of the extremities but we have noticed an enlargement of the ears with increased vascularity in the heat-acclimated animals.

N. KOSOWER: You mentioned that your cells might be analagous in a way to hereditary spherocytosis. Would you care to comment on this? I recall some studies by Motulsky who has, I think, shown that when you do increase the temperature in spherocytosis of mice they are even more susceptible to hemolysis.

N. MEYERSTEIN: Hereditary spherocytosis cells have some features in common with the red cells of heat-acclimated hamsters, such as increased glucose uptake and osmotic fragility, although the latter cells have a low ATP level, normal mean corpuscular hemoglobin concentration and they are not spherocytes.

R. JIJI (*USA*): Did you study the ouabain sensitivity in relation to hereditary spherocytosis?

N. MEYERSTEIN: We have not yet done ouabain sensitivity studies on these cells.

B. RAMOT: My comment is that you cannot compare a model system to human diseases, such as congenital spherocytosis or PK deficiency, since increased osmotic fragility is a very unspecific finding.

N. MEYERSTEIN: It is true that this is a model system of red cells which have been transformed by heat acclimation. A model which, however, has certain features in common with human disease can teach us about the mode of action of the disease, because of their similarities and differences.

B. RAMOT: An increased red cell fragility is found in so many situations that I don't think you can make a diagnosis from such a finding only.

The papers by Drs. Danon and Kleihauer are now open for discussion.

A. BEN-DAVID (*Israel*): Dr. Danon, in some of your pictures the labeling of the membrane of the adult and the old red cells by colloid iron particles seem to be higher than in the reticulocytes. I wonder whether this is a technical point and you are comparing ratios or whether there is a real difference, namely, that the adult red cell has a higher negative charge than the reticulocyte. If so, why isn't the reticulocyte swallowed?

D. DANON: You are right, the reticulocyte is less negatively charged than the adult red cell and this is not a new observation; it was described by Stephens in 1941. It has also been thoroughly studied by Dr. Doljansky (Jerusalem). We have studied the sequence of events from the earliest recognizable erythroid cell. After every division there is a reduction in surface charge until the nucleus is expelled with a minimum charge on its membrane, leaving behind a reticulocyte which is less negatively charged than the adult red cell. As the reticulocyte shrinks, its surface to volume ratio becomes smaller. When it becomes a young red cell it reaches its maximum charge and then it loses charge gradually until it becomes old. Dr. Ramot will tell you that in certain anemias reticulocytes are removed from the circulation. These reticulocytes have a charge similar to that found in old cells.

G. IZAK (*Israel*): How can you explain that when erythroid nuclei are expelled they take along with them part of the membrane? How is it that the number of colloidal particles on that membrane is so much smaller than on the membrane they leave behind? What happens to those cells that lose their nuclei not by expulsion, but by intracellular break-up? If I may continue with the last question to Dr. Kleihauer, I should like to ask him to comment further or to speculate on that very interesting observation of the extra-corpuscular factor producing Heinz bodies in adult cells.

D. DANON: To your first question, you gave the answer yourself—part of the membrane leaves with the nucleus. During nuclear expulsion a new membrane is apparently formed, as during mitosis. The synthesis of new membrane during division is apparently not accompanied by the formation of additional N-acetyl-neuraminic acid for the new surface. Every cell division is followed by less charge on the daughter cells. This is a tentative explanation; we still do not have the proof that there is no synthesis of N-acetyl neuraminic acid in the membrane that is synthesized during nuclear expulsion. To your second question the answer is that I do not know of such nuclei that break up within the cell.

E. KLEIHAUER: I think it is very difficult to answer the question of Dr. Izak. I will try to recall some facts and speculate on them. Heinz body formation depends on the presence of hemoglobin but the high susceptibility of red cells of newborn infants is independent of the fetal or adult type of hemoglobin. Furthermore, it is unlikely that the decreased activities of some red cell enzymes are responsible for it. Let us also assume that acetylphenylhydrazine is activated in the same way by adult and cord blood, so we are left with the membrane. If it is not the membrane, we have to look for other extra-corpuscular factors which may cause the enhanced Heinz body formation. Since acetylphenylhydrazine is used to produce Heinz bodies, a difference in the activation of acetylphenylhydrazine by extra-corpuscular factors present in the blood of adults and newborns has to be considered.

J. MAGER: I want to ask Dr. Danon a naive question. How does one survive an attack of influenza, in view of the fact that the neuraminidase-treated cells have such a reduced life span and since influenza virus contains a very high content of receptor-destroying enzyme activity?

D. DANON: If you incubate red cells with influenza virus *in vitro* until the surface charge is markedly reduced and then inject these cells back into the circulation, they will not survive for long. However, I do not know how much of this receptor-destroying enzyme activity will take place *in vivo* in the circulation. If it does, we should consider the number of influenza viruses in the circulation as compared to the number of cells. It might be a question of dose. I believe that anemia is rare in influenza. There is, however, a loss of a certain number of cells during influenza.

E. KLEIHAUER: We could not observe enhanced Heinz body formation in adult red cells when suspended in plasma of newborn infants over a period of 15 min before acetylphenylhydrazine was added. It is possible that the incubation time was too short in order to get an exchange between phospholipids of plasma and membrane. On the other hand, an extended incubation time might possibly produce alterations in red cell properties so that it would be difficult to come to final conclusions from the mentioned experiments.

J. C. KAPLAN: I have a question for Dr. Danon and a question for Dr. Kleihauer. To Dr. Danon it is rather a statement. I remember that in Dr. Bessis' laboratory in Paris, I saw a fantastic film showing how clever macrophages are. When a single red cell is shot with a laser beam, a very minute lesion of the membrane is created, and at once the macrophages rush to ingest this cell. I am just mentioning that because it might be of interest in the discussion.

I would like to ask Dr. Kleihauer how the experiment with APH was carried out. The original Beutler test is done with whole blood. Do you think that you would get the same result with washed blood?

E. KLEIHAUER: We have also done these investigations with washed red cells suspended in buffered saline solution. Heinz body production by acetylphenylhydrazine remained different between the fetal and adult red cells.

D. DANON: Bessis also showed in 1952, on phase contrast microcinematography, the expulsion of the nucleus from the normoblast *in vitro*. He did not show whether the nucleus was surrounded by a membrane or not. The macrophages are capable of recognizing an experimentally damaged cell; one of the beautiful demonstrations is that of Bessis when he uses a laser beam of 1μ. This beam destroys the red cell but we do not know what happens to the membrane. I would like to ask a question, Dr. Kleihauer. Did you try to submit the adult cells to the physical trauma by passing them through a syringe and needle several times as you do in transfusion?

R. SCHNEIDER (*USA*): In connection with a study of erythrocytes containing the unstable hemoglobin Sabine, we found that dark field microscopy, especially with blue light, provides improved visualization of intraerythrocytic inclusions. Fig. 1 shows smears, made in dark field microscopy with blue light, of heparinized normal adult blood both untreated and incubated at 37 C. We now make the smears with saline-washed erythrocytes, in order to eliminate artifacts due to reflections of light from the serum proteins. In the untreated sample, the erythrocytes are homogeneous grey, except for a few yellowish green, punctate inclusions. These increase progressively on incubation at 37 C; by 24 hr there are usually several that resemble Heinz bodies and by 48 hr a few inclusions appear to have enlarged and coalesced so that they completely fill a cell.

Fig. 2 is of a cord blood sample under similar conditions. In the untreated sample the intraerythrocytic inclusions are considerably more prominent than in the adult, and on incubation they become even more so.

Fig. 3 is of a smear of untreated heparinized blood of a sickle cell anemia patient. The same microscopic darkfield preparation is seen with blue light (top) and with tungsten light (bottom). In sickle cell anemia, there are intraerythrocytic inclusions of varying size and shape in many, often most, of the cells. Defects of the erythrocyte membrane and partially or completely extruded particles are also demonstrable. Using this technique we find inclusion bodies in most cord blood samples and also in some hemoglobinopathies.

L. KLEJAN (*Israel*): I should like to ask Dr. Kleihauer, whether experiments with labeled APH could produce an answer as to whether the fetal cells are more permeable to this compound.

Discussion

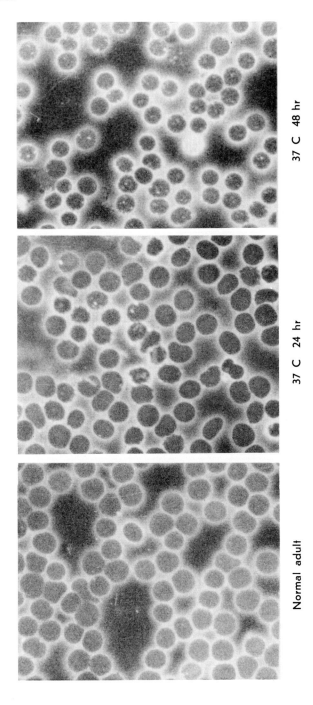

FIG. 1 Dark field microscopy of heparinized adult blood incubated at 37 C for 24 and 48 hr.

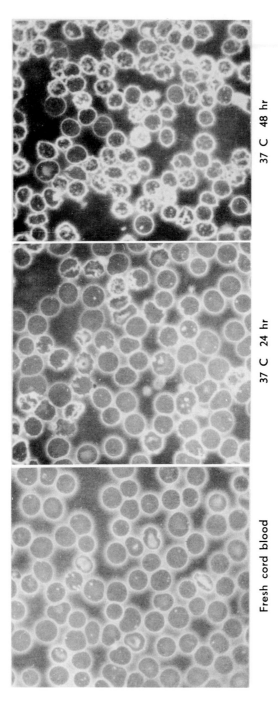

FIG. 2 Dark field microscopy of heparinized cord blood incubated at 37 C for 24 and 48 hr.

Discussion

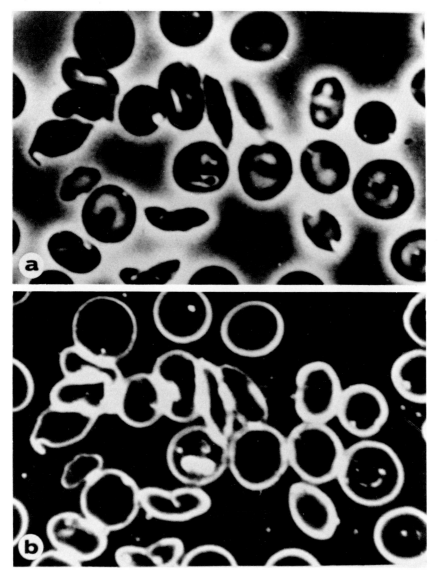

FIG. 3 Smears of heparinized blood of sickle cell anemia patient, in darkfield microscopy; (a) with blue light, (b) same field, tungsten light.

E. KLEIHAUER: There have been frequent discussions in our laboratory about performing these experiments. Labeling of APH would give us additional information not only about the penetration but also about the action of APH. To my knowledge it is not known whether APH is deacetylated before it penetrates the membrane. Double labeling of APH could clarify this question.

E. M. KOSOWER: It is really unfair for a chemist to have the last word, but from what has been done, I would say that phenylhydrazine is acylated in order to become active for the oxidation of GSH and not the other way around. Phenylhydrazine, as one of its reactions in biological systems, reacts with glucose to form phenylhydrazones, perhaps leading to further oxidation of the glucose and then a lot of the phenylhydrazine is excreted in the form of acyl derivatives of sugars. I think that the phenylhydrazine is probably converted into an acyl derivative and APH is the precursor of the active form, the oxidation product of APH. I would really recommend to anyone who wants to do stoicheiometric studies on GSH oxidation followed by subsequent free radical generation, to use either azoester or perhaps utilize the oxidation product of APH. Azoester is commercially available from Cal-Biochem (USA). The oxidation product of APH has not been studied. The APH oxidation product would produce phenyl radicals in exactly the same way as azoesters. Enough chemistry has been done on that kind of compound to show that in spite of the fact that they look like different compounds, the chemistry would be exactly parallel. Dr. Schneider, could I ask you if you incubated your cells with anything?

R. SCHNEIDER: No. These incubations were performed without additives.

SESSION II

Chairmen: N. Kosower, *USA*
I. Izak, *Israel*

Participants: G. Bianco, *Italy*
E. Gallo, *Italy*
U. Mazza, *Italy*
V. Prato, *Italy*
E. A. Rachmilewitz, *Israel*
H. M. Ranney, *USA*
G. Ricco, *Italy*
R. F. Rieder, *USA*

ASPECTS OF THE STRUCTURE, SYNTHESIS AND CLINICAL EFFECTS OF UNSTABLE HEMOGLOBINS

STUDIES ON HEMOGLOBINS ZÜRICH, GUN HILL AND PHILLY

RONALD F. RIEDER[*]

Department of Medicine, State University of New York, Downstate Medical Center, Brooklyn, New York, USA

Since the initial description of Hb S (1) more than 100 mutant forms of human hemoglobin have been discovered (2). Most of these variants produce no clinical manifestations (3). Hemolytic anemia is characteristic of patients possessing Hb S, Hb C, Hb D Punjab or Hb E in the homozygous state or in the heterozygous state when in association with another abnormal hemoglobin or thalassemia (4). Recently it has become evident that there is a class of unstable abnormal hemoglobins capable of causing varying degrees of hemolytic disease in patients heterozygous for the gene for the abnormal protein (5, 6).

Development of the evidence for the class of unstable hemoglobins. Since the 19th century it has been known that exposure of normal individuals to certain organic chemicals related to phenylhydrazine can result in acute hemolytic anemia associated with the appearance of intraerythrocytic inclusions or Heinz bodies (7). These inclusion bodies are generally best visualized with supravital stains and are often accompanied by the presence in the blood of methemoglobin and a poorly defined group of hemochromes termed sulfhemoglobin or verdoglobin (8). The common action of the offending chemicals appears to involve an oxidative denaturation of hemoglobin, causing precipitation of the protein in the form of Heinz bodies (9, 10). The erythrocytes of patients with a deficiency of the red blood cell enzyme glucose-6-phosphate dehydrogenase are inordinately subject to denaturation of their hemoglobin and can develop hemolytic

---

[*] Career Scientist, Health Research Council of the City of New York (I-633).

episodes when ingesting primaquine and related compounds, which generally exert no deleterious effect on normal erythrocytes (11).

In 1952, Cathie reported an infant with an apparently idiopathic form of Heinz body anemia (12). Subsequently a number of similar cases of unexplained congenital Heinz body anemia were discovered, some of which were associated with the excretion in the urine of a dark pigment, classed as a bilifuscin (13–17).

The first case of Heinz body anemia with definite evidence of an associated abnormal hemoglobin was described by Scott and co-workers in 1960 (18). The patient, a Caucasian boy, had a history of hemolytic anemia since the age of two. Splenectomy had been performed at 32 months of age. When examined at age 13 years, 90% of the peripheral red blood cells contained Heinz bodies. Hemoglobin electrophoresis showed a poorly defined smear trailing behind the main Hb A band but there was no further characterization of the abnormal fraction.

The discovery and chemical analysis of Hb Zürich provided the first example of a Heinz body anemia due to a well defined hemoglobin mutant (19, 20). Fifteen members of four generations of a Swiss family were found to have an abnormal hemoglobin with an electrophoretic mobility similar to Hb S. Two patients developed severe hemolytic episodes upon ingestion of sulfonamides. *In vivo* studies indicated that a variety of sulfonamide agents and oxyquinoline drugs were capable of causing hemolytic episodes in patients with Hb Zürich. These episodes were associated with the appearance of erythrocytic inclusion bodies. No inclusions were present in the red blood cells during periods when drugs were not administered. During the acute hemolytic periods, the red cell half-life was reduced markedly to one and one-half days. In the basal state, the red cell half-life of patients with Hb Zürich was shortened to 11 days.

The abnormal hemoglobin was more heat labile and had a significantly higher rate of methemoglobin formation than Hb A. Fingerprint and amino acid analyses revealed that in Hb Zürich the distal histidine in contact with the heme group at position 63 of the β-chain was replaced by arginine (21).

Since the description of Hb Zürich, more than 20 different unstable hemoglobins associated with hemolytic states have been discovered (5, 6).

Experience with three unstable hemoglobins: Hb Zürich, Hb Gun Hill and Hb Philly. A second family with Hb Zürich was found in Maryland. The members appeared to be of North European origin but a definite nationality of the ancestors could not be ascertained. Six individuals from three

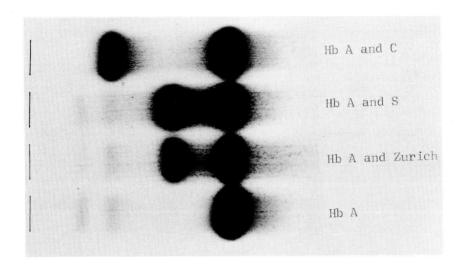

FIG. 1. Hemoglobin electrophoresis on starch gel at pH 8.6. Hb Zürich has a mobility similar to Hb S. The anode is on the right. (Reprinted with permission from ref. 22.)

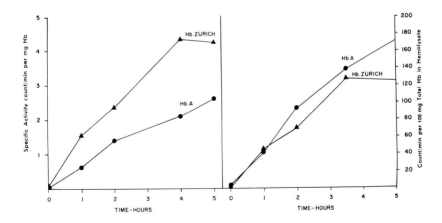

FIG. 2. Incorporation of Fe^{59} into Hb A and Hb Zürich by reticulocytes *in vitro*. The graph on the left shows the specific activities; that on the right shows the proportion of total counts incorporated into each fraction. (Reprinted with permission from ref. 22.)

generations were found to be heterozygous for the abnormal hemoglobin. Fig. 1 shows the starch gel electrophoretic pattern of Hb Zürich compared to Hb A, Hb S and Hb C.

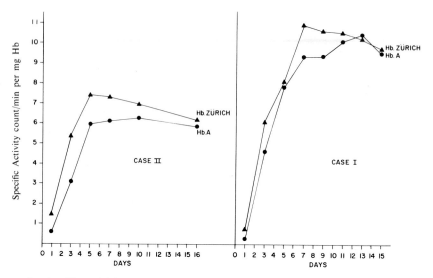

FIG. 3. Specific activities of Hb Zürich and Hb A following the i.v. administration of Fe^{59} to two patients heterozygous for the mutant hemoglobin. (Reprinted with permission from ref. 22.)

The increased tendency of Hb Zürich to denature and precipitate when exposed to mild oxidant stress was illustrated by the formation of multiple Heinz-like inclusion bodies when the erythrocytes were incubated with brilliant cresyl blue. When hemolysates containing Hb Zürich were mixed with solutions of brilliant cresyl blue, amorphous precipitates formed which were followed by the development of coccoid bodies resembling the intra-erythrocytic inclusions.

The proportion of Hb Zürich in the hemolysates ranged from 19 to 27% of the total hemoglobin. However, the results of *in vivo* and *in vitro* studies of hemoglobin synthesis suggested that Hb A and Hb Zürich were actually made in approximately equal amounts. When reticulocytes were incubated with Fe^{59}, the specific activity of Hb Zürich was always greater than that of Hb A. Total incorporation of radioactivity into the two fractions was approximately equal (Fig. 2).

Two patients were given Fe^{59} i.v. and incorporation into Hb A and Hb Zürich was measured. The specific activity of Hb Zürich was greater than that of Hb A for a period of one week (Fig. 3). After 10 days, the specific activities of the two hemoglobin fractions were approximately equal. Twenty-four hr after injection, the total radioactivity in Hb Zürich equaled that in Hb A.

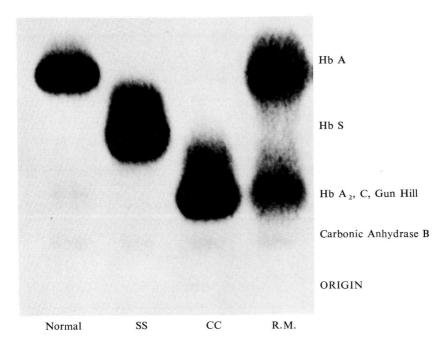

FIG. 4. Starch gel electrophoresis at pH 8.6. Hb Gun Hill (patient R.M.) has a mobility similar to Hb C. The anode is towards the top.

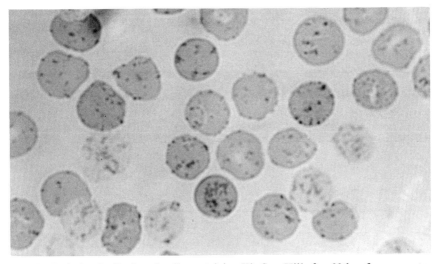

FIG. 5. Inclusion bodies in red cells containing Hb Gun Hill after 22 hr of exposure to brilliant cresyl blue. Normal red cells showed only an occasional inclusion body after similar treatment.

The results of the studies of hemoglobin synthesis suggest that the two hemoglobins are made in equal amounts and that the final unequal percentages in the peripheral blood are due to preferential destruction of the labile Hb Zürich.

Hb Gun Hill was discovered in a man of German-English ancestry with a long history of mild scleral icterus (5, 23). One daughter also had the abnormal hemoglobin. The propositus had a slightly enlarged spleen but had never had deep jaundice, anemia or dark urine. The reticulocyte count was 4 to 6% and the erythrocyte half-life was reduced to 17 days.

On starch gel electrophoresis (Fig. 4), Hb Gun Hill had a mobility similar to Hb C or Hb A_2. The heat stability of Hb Gun Hill was decreased when compared to Hb A. Incubation of erythrocytes with brilliant cresyl blue caused the formation of inclusion bodies (Fig. 5).

Fingerprint and amino acid analysis revealed that in the β-chains of Hb Gun Hill there is a five amino acid deletion which includes the heme-

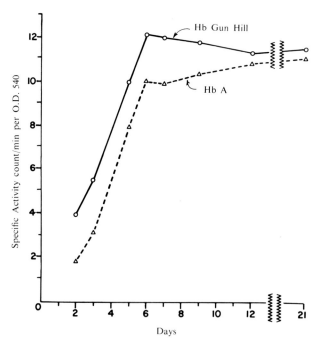

FIG. 6. Specific activities of Hb Gun Hill and Hb A following i.v. administration of Fe^{59}. (Reprinted from *Rieder, R.F.* and *Bradley, T.B., Jr.* Hemoglobin Gun Hill: An unstable protein associated with chronic hemolysis. *Blood* **32**: 355, 1968.)

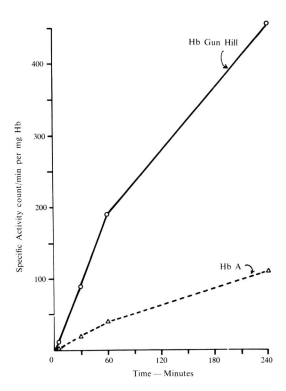

FIG. 7. Incorporation of C^{14}-leucine into Hb Gun Hill and Hb A by reticulocytes *in vitro*. (Reprinted from *Rieder, R.F.* and *Bradley, T.B., Jr.* Hemoglobin Gun Hill: An unstable protein associated with chronic hemolysis. *Blood* **32**: 355, 1968.)

binding proximal histidine at position 92 (23). Hb Gun Hill lacks heme groups on the β-chains.

In patients with Hb Gun Hill, 30 to 35% of the hemoglobin in the peripheral blood consisted of the abnormal fraction. However, following administration of Fe^{59} to the propositus, the specific activity of the mutant hemoglobin was greater than that of Hb A (Fig. 6). The data indicate that approximately equal amounts of newly synthesized Hb A and Hb Gun Hill were present in the peripheral blood two days after isotope administration. These results are similar to those obtained in the experiments with Hb Zürich and suggest unequal turnover rates for Hb A and Hb Gun Hill.

In vitro studies of hemoglobin synthesis utilizing C^{14}-leucine produced unusual results (Fig. 7). The specific activity of Hb Gun Hill was four times greater than that of Hb A and there was actually almost twice as

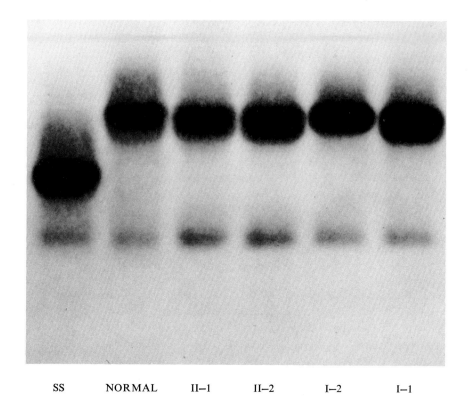

FIG. 8. Starch gel electrophoresis at pH 8.6 of hemolysates prepared from the blood of four members of the family in which Hb Philly was found (I-1, father; I-2, mother; II-1, propositus; II-2, brother). No abnormal hemoglobin band is noted. The anode is towards the top of the Fig.

much total radioactivity incorporated into the mutant fraction. Further studies of this phenomenon, which suggest more rapid synthesis of Hb Gun Hill than Hb A, are in progress.

A third abnormal unstable hemoglobin that we have studied is Hb Philly (24). This mutant was found in an eight-year-old girl of European ancestry and in two members of her family. The only significant clinical findings were an inconstant slight splenomegaly and a 5 to 8% reticulocytosis.

Heating of the hemolysate of the propositus to 50 C produced hemoglobin precipitation, and incubation of her red blood cells with brilliant cresyl blue caused formation of inclusion bodies. However, no abnormal hemoglobin was detected on starch gel electrophoresis (Fig. 8). Similarly,

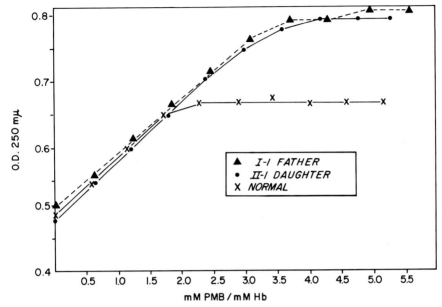

FIG. 9. PMB titration of hemolysates from the propositus with Hb Philly and her father compared with a normal hemolysate. In the normal, the exhaustion of free sulfhydryl groups, indicated by the break in the curve, occurs at a ratio of two molecules of PMB per molecule of hemoglobin. Both hemolysates from the family members show the change of slope of the titration curve at a point where four molecules of sulfhydryl group have reacted per molecule of hemoglobin. (Reprinted with permission from ref. 24.)

column chromatography on a variety of media failed to demonstrate an abnormal hemoglobin fraction.

Titration of the hemolysate of the propositus with paramercuribenzoate (PMB) indicated that there was an average of four reactive sulfhydryl groups per hemoglobin molecule in place of the normal two groups (Fig. 9). Starch gel electrophoresis of the PMB titrated hemolysate revealed that approximately 35% of the hemoglobin had been split into α- and β-chains (Fig. 10). Normal hemolysates showed no splitting under the conditions used.

The PMB titration provided evidence for the presence of an abnormal hemoglobin unusually susceptible to dissociation into subunits. The splitting of the abnormal hemoglobin by PMB allowed the chromatographic separation of the split α- and β-chains of Hb Philly from the Hb A fraction which remained intact.

In Hb Philly there is a substitution of phenylalanine for tyrosine at

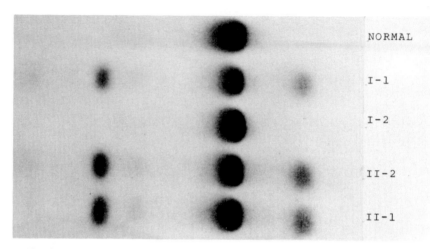

FIG. 10. Starch gel electrophoresis of PMB treated hemolysates from four members of the family with Hb Philly compared with a normal. The anode is to the right of the Fig. Hemolysates from all family members except the mother (I-2) show the appearance of free β-chains (spot anodal to Hb A) and free α-chains (spot cathodal to Hb A) after exposure to PMB. (Reprinted with permission from ref. 24.)

position 35 of the β-chain. In Hb A, the tyrosine at β 35 is situated at one of the contacts between the α- and β-chains and forms a hydrogen bond with aspartic acid at position α 126 (2). The substitution of phenylalanine for tyrosine in Hb Philly prevents formation of this hydrogen bond with resultant weakening of the forces which hold the hemoglobin tetramer together. As a result, the normally buried sulfhydryls at β 112 and α 104 are exposed and subject to titration by PMB.

DISCUSSION

The severity of the hemolytic disorders which occur in patients with unstable hemoglobins is quite variable. Hb Philly and Hb Gun Hill cause only mild compensated hemolysis while Hb Hammersmith (25) and Hb Köln (26) produce a more severe anemia. Patients with Hb Zürich developed anemia only when challenged with oxidant drugs although there was evidence of constant mildly increased hemolysis.

The appearance of large numbers of Heinz bodies in the peripheral blood of patients with these disorders seems to be a function of splenectomy. The first appearance of the inclusion bodies shortly after removal of the spleen has been documented several times (17, 26–29). Crosby has suggested that the intact spleen is capable of removing Heinz bodies from circulating erythrocytes, after which process the cells may be returned to the circulation (30). Splenectomy would thus destroy this clearance mechanism allowing the appearance of Heinz bodies in large numbers. The patients with Hb Gun Hill and Hb Philly have not been splenectomized and do not have spontaneous red cell inclusion bodies. Patients with Hb Zürich have exhibited Heinz bodies only during the course of acute hemolytic episodes following drug treatment.

Heinz body-like inclusions were produced in the cells of patients with Hb Zürich, Hb Gun Hill and Hb Philly by incubation with a redox dye such as brilliant cresyl blue or new methylene blue. This procedure appears to be a good screening test for the presence of an unstable hemoglobin although at least one unstable mutant, Hb Hasharon, did not readily form inclusions when red blood cells were exposed to the dyes (31). The precipitation of solutions of Hb Zürich by brilliant cresyl blue suggests that the intracellular inclusions represent denatured precipitated hemoglobin. Studies of the similar inclusions found in thalassemia indicate that they are formed from precipitated hemoglobin (32).

Increased hemoglobin precipitation upon heating of hemolysates to 50 C was observed with Hb Zürich, Hb Gun Hill and Hb Philly and has been reported in studies on other unstable hemoglobins (27–29, 33–36). This procedure also has value in the initial screening of patients with unexplained hemolysis.

The amino acid substitutions observed in the unstable hemoglobins have generally involved the substitution of one uncharged amino acid for another (2). Thus several of the mutants have not been detectable by electrophoresis (25, 37, 38). Hb Philly could not be separated from Hb A until it was preferentially split into α- and β-monomers by PMB treatment. It appears likely that several of the early reports of idiopathic congenital Heinz body anemia may have included examples of unstable hemoglobins. The difficulty in readily separating these mutants from Hb A by routine electrophoretic techniques stresses the value of the heat and brilliant cresyl blue tests as detection procedures.

Hb Gun Hill (23) and Hb Sabine (36) have electrophoretic mobilities which differ from that of Hb A. However, globin preparations failed to show abnormally charged α- or β-chains when examined by starch gel electrophoresis in urea containing buffers. The electrophoretic differences from normal hemoglobin observed in these two mutants probably results from loss of heme groups from the abnormal polypeptide chains. Heme deficits have been demonstrated in both these proteins and in Hb Köln (39). Heme loss from the unstable mutants may be the cause of the pigmented urine (dipyrroluria) found in certain patients with Heinz body hemolytic anemia.

The amino acid substitutions causing instability of the hemoglobin tetramer have been observed at nonpolar positions in contact with the heme groups, at general internal positions, and at the contacts between the subunit polypeptide chains. The net effect of any of these mutations can be explained as a weakening of the forces which either a) tend to bind the porphyrin groups to the protein, b) bind the protein subunits to each other or c) determine important configurational characteristics of the individual chains (2). Hemoglobins having defects in the binding of porphyrin to protein include Zürich, Gun Hill, Köln, Hammersmith, Torino and Santa Ana (2). Although several hemoglobins have been reported with mutations at the interchain contacts, Hb Philly appears to be the only hemoglobin with faulty binding of protein subunits to each other with clinical effects resulting from instability of the molecule. Hb Genova and Hb Wien are two hemoglobins associated with substitutions at general internal positions and increased red blood cell destruction (2). The greater frequency and severity of the effects of alterations in heme binding may be a reflection of the marked instability of heme-free globin.

The studies on the synthesis of Hb Zürich and Hb Gun Hill indicate that the relative proportions of these hemoglobins and Hb A in the peripheral blood give no indication of the relative synthetic rates. The unstable hemoglobins have a more rapid turnover than normal hemoglobin. The mechanism for this rapid turnover may involve either preferential destruction of cells containing largest amounts of a heterogeneously distributed mutant protein or selective removal of denatured hemoglobin from intact circulating cells. At present there is no evidence to favor either mechanism.

The destruction of red blood cells containing easily denatured hemoglobin may be due to a deleterious effect of the inclusion bodies on the erythrocyte membrane. The formation of mixed disulphide bonds between Heinz bodies and the red cell membrane has been proposed as a mechanism which might lead to increased hemolysis (40).

Original studies reviewed in this article were performed in association with Drs. W. H. Zinkham and N. A. Holtzman (Hb Zürich), T. B. Bradley, Jr. (Hb Gun Hill), F. A. Oski and J. B. Clegg (Hb Philly).

Supported by U.S.P.H.S. Research Grant AM-12401.

* *Note added in proof:* Recently, it has been demonstrated that in reticulocytes containing unstable β-chain mutant hemoglobins, there is a free pool of α-chains. [WHITE, J. M. and BRAIN, M. C. Defective synthesis of an unstable haemoglobin: Hemoglobin Koln (β^{98} Val-Met). *Brit. J. Haemat.* **18**: 195, 1970; RIEDER, R. F. Synthesis of hemoglobin Gun Hill: Increased synthesis of the heme-free β^{GH} globin chain and subunit exchange with a free α-chain pool. Submitted for publication.] Direct intracellular exchange of α subunits has been demonstrated between the free pool and preformed Hb Gun Hill. This results in an increased specific activity of the unstable hemoglobin above that due to synthesis of new abnormal β-chains. Direct measurement of synthesis of individual polypeptide chains has confirmed the finding that synthesis of β^{GH} is greater than that of β^A.

REFERENCES

1. PAULING, L., ITANO, H. A., SINGER, S. J. and WELLS, I. C. Sickle cell anemia, a molecular disease. *Science* **110**: 543, 1949.
2. PERUTZ, M. F. and LEHMANN, H. Molecular pathology of human hemoglobin. *Nature (Lond.)* **219**: 902, 1968.
3. HELLER, P. Hemoglobinopathic dysfunction of the red cell. *Amer. J. Med.* **41**: 799, 1966.
4. WINTROBE, M. M. "Clinical hematology," 6th edn. Philadelphia, Lea and Febiger, 1967.
5. RIEDER, R. F. and BRADLEY, T. B., JR. Hemoglobin Gun Hill: An unstable protein associated with chronic hemolysis. *Blood* **32**: 355, 1968.
6. CARRELL, R. W. and LEHMANN, H. The unstable haemoglobin haemolytic anaemias. *Seminars Hemat.* **6**: 116, 1969.
7. HEINZ, R. Morphologische Veränderungen der rothen Blutkörperchen durch Gifte. *Virchows Arch. path. Anat.* **122**: 112, 1890.
8. FERTMAN, M. H. and FERTMAN, M. B. Toxic anemias and Heinz bodies. *Medicine (Baltimore)* **34**: 131, 1955.
9. JANDLE, J. H., ENGLE, L. K. and ALLEN, D. W. Oxidative hemolysis and precipitation of hemoglobin. I. Heinz body anemias as an acceleration of red cell aging. *J. clin. Invest.* **39**: 1818, 1960.
10. ALLEN, D. W. and JANDLE, J. H. Oxidative hemolysis and precipitation of hemoglobin. II. Role of thiols in oxidant drug action. *J. clin. Invest.* **40**: 454, 1961.
11. BEUTLER, E. The hemolytic effect of primaquine and related compounds. A review. *Blood* **14**: 103, 1959.
12. CATHIE, I. A. B. Apparent idiopathic Heinz body anaemia. *Gt. Ormond Str. J.* **3**: 43, 1952.
13. LANGE, R. D. and AKEROYD, J. H. Congenital hemolytic anemia with abnormal pigment metabolism and red cell inclusion bodies: A new clinical syndrome. *Blood* **13**: 950, 1958.
14. SCHMID, R., BRECHER, G. and CLEMENS, T. Familial hemolytic anemia with erythrocytic inclusion bodies and a defect in pigment metabolism. *Blood* **14**: 991, 1959.
15. WORMS, R., BERNARD, J., BESSIS, M. and MALASSENET, R. Anemia hemolytique congenitale avec inclusions intra-erythrocytaires et urines noires. Rapport d'un nouveau cas avec etude de microscopie electronique. *Nouv. Rev. franç. Hémat.* **1**: 805, 1961.
16. LELONG, M., FLEURY, J., ALAGILLE, D., MALASSENET, R., LORTHOLARY, P. and PARA, M. L'anemie hemolytique constitutionelle nonspherocytaire avec pigmenturie. Un cas avec étude enzymatique. *Nouv. Rev. franç. Hémat..* **1**: 819, 1961.
17. MOZZICONACCI, P., ATTAL, C., PHAM-HUU-TRUNG, MALASSENET, R. and BESSIS, M. Nouveau cas d'anemie hemolytique congenitale avec inclusions intra-erythrocytaires et urines noires. Importance de la splenectomie dans l'apparition de ces inclusions. *Nouv. Rev. franç. Hémat.* **1**: 832, 1961.

18. SCOTT, J. L., HAUT, A., CARTWRIGHT, G. E. and WINTROBE, M. M. Congenital hemolytic disease associated with red cell inclusion bodies, abnormal pigment metabolism and an electrophoretic hemoglobin abnormality. *Blood* **16**: 1239, 1960.
19. FRICK, P. G., HITZIG, W. H. and BETKE, K. Hemoglobin Zürich. I. A new hemoglobin anomaly associated with acute hemolytic episodes with inclusion bodies after sulfonamide therapy. *Blood* **20**: 261, 1962.
20. BACHMANN, F. and MARTI, H. R. Hemoglobin Zürich. II. Physicochemical properties of the abnormal hemoglobin. *Blood* **20**: 272, 1962.
21. MULLER, C. J. and KINGMA, S. Haemoglobin Zürich, $\alpha_2^A\beta_2$ 63 Arg. *Biochim. biophys. Acta (Amst.)* **50**: 595, 1961.
22. RIEDER, R. F., ZINKHAM, W. H. and HOLTZMAN, N. A. Hemoglobin Zürich. Clinical, chemical and kinetic studies. *Amer. J. Med.* **39**: 4, 1965.
23. BRADLEY, T. B., JR., WOHL, R. C. and RIEDER, R. F. Hemoglobin Gun Hill: deletion of five amino acid residues and impaired heme-globin binding. *Science* **157**: 1581, 1967.
24. RIEDER, R. F., OSKI, F. A. and CLEGG, J. B. Hemoglobin Philly (β35 Tyrosine → Phenylalanine): Studies in the molecular pathology of hemoglobin. *J. clin. Invest.* **48**: 1627, 1969.
25. DACIE, J. V., SHINTON, N. K., GAFFNEY, P. J., CARRELL, R. W. and LEHMANN, H. Haemoglobin Hammersmith (β42 (CD1) Phe → Ser). *Nature (Lond.)* **216**: 663, 1967.
26. HUTCHISON, H. E., PINKERTON, P. H., WATERS, P., DOUGLAS, A. S., LEHMANN, H. and BEALE, D. Hereditary Heinz-body anaemia thrombocytopenia, and haemoglobinopathy (Hb Köln) in a Glasgow family. *Brit. med. J.* **2**: 1900, 1964.
27. GRIMES, A. J., MEISLER, A. and DACIE, J. V. Congenital Heinz-body anemia. Further evidence on the cause of Heinz-body production in red cells. *Brit. J. Haemat.* **10**: 281, 1964.
28. DACIE, J. V., GRIMES, A. J., MEISLER, A., STEINGOLD, L., HEMSTED, E. H., BEAVEN, G. H. and WHITE, J. C. Hereditary Heinz-body anaemia. A report of studies of five patients with mild anaemias. *Brit. J. Haemat.* **10**: 388, 1964.
29. OPFELL, R. W., LORKIN, P. A. and LEHMANN, H. Hereditary non-spherocytic haemolytic anaemia with post-splenectomy inclusion bodies and pigmenturia caused by an unstable haemoglobin. Santa Ana-β88 (F4) leucine → proline. *J. Med. Genet* **5**: 292, 1968.
30. CROSBY, W. H. Normal functions of the spleen relative to red blood cells: a review. *Blood* **14**: 399, 1959.
31. CHARACHE, S., MONDZAC, A. M. and GESSNER, U. Hemoglobin Hasharon (α_2 47 his (CD5) β_2): A hemoglobin found in low concentration. *J. clin. Invest.* **48**: 834, 1969.
32. FESSAS, P., LOUKOPOULOS, D. and KALTSOYA, A. Peptide analysis of the inclusions of erythroid cells in β-thalassemia. *Biochim. biophys. Acta (Amst.)* **124**: 430, 1966.
33. VAUGHAN JONES, R., GRIMES, A. J., CARRELL, R. W. and LEHMANN, H. Köln haemoglobinopathy. Further data and a comparison with other hereditary Heinz-body anaemias. *Brit. J. Haemat.* **13**: 394, 1967.
34. RANNEY, H. M., JACOBS, A. S., UDEM, L. and ZALUSKY, R. Hemoglobin Riverdale-Bronx. An unstable hemoglobin resulting from the substitution of arginine for glycine at helical residue B6 of the β polypeptide chain. *Biochem. biophys. Res. Commun.* **33**: 1004, 1968.
35. SATHIAPALAN, R. and ROBINSON, M. G., Hereditary haemolytic anaemia due to an abnormal haemoglobin (haemoglobin Kings County). *Brit. J. Haemat.* **15**: 579, 1968.
36. SCHNEIDER, R. G., UEDA, S., ALPERIN, J. B., BRIMHALL, B. and JONES, R. T. Hemoglobin Sabine β91 (F7) Leu → Pro. An unstable variant causing severe anemia with inclusion bodies. *New Engl. J. Med.* **280**: 739, 1969.
37. SANSONE, G., CARRELL, R. W. and LEHMANN, H. Haemoglobin Genova: β28 (B10) Leucine → Proline. *Nature (Lond.)* **214**: 877, 1967.
38. BERETTA, A., PRATO, V., GALLO, E. and LEHMANN, H. Haemoglobin Torino-α43 (CD1) Phenylalanine → Valine. *Nature (Lond.)* **217**: 1016, 1968.

39. JACOB, H. S., BRAIN, M. C., DACIE, J. V., CARRELL, R. W. and LEHMANN, H. Abnormal haem binding and globin SH group blockade in unstable haemoglobins. *Nature (Lond.)* **218**: 1214, 1968.
40. JACOB, H. S., BRAIN, M. C. and DACIE, J. V. Altered sulfhydryl reactivity of hemoglobins and red blood cell membranes in congenital Heinz body hemolytic anemia. *J. clin. Invest.* **47**: 2664, 1968.

HEMOGLOBINS WITH ALTERED OXYGEN AFFINITY

HELEN M. RANNEY

Department of Medicine, Albert Einstein College of Medicine, Bronx, New York

Since the structure of hemoglobin may be assumed to represent the result of selective pressures favoring an ideal oxygen transport protein, the replacement of certain amino acid residues in human hemoglobin variants would be expected to result in altered reactions of the variant hemoglobin with oxygen. In only a small proportion of the known variants are the relationships of the abnormal oxygen equilibria to the accompanying clinical syndromes readily evident. Hemoglobins with increased oxygen affinity may be associated with erythrocytosis, and hemoglobins in which half of the hemes are oxidized (the M hemoglobins) or in which the oxygen affinity is low may be associated clinically with cyanosis. However, the number of amino acid substitutions which result in hemoglobins of high oxygen affinity cannot be deduced from the clinical association with erythrocytosis, since many of the hemoglobin variants exhibiting high oxygen affinity are also quite unstable and the affected individuals present with a Heinz body hemolytic anemia rather than erythrocytosis. Similarly the clinical syndrome accompanying the presence of Hb H, (β_4^A) which has a very high oxygen affinity, is that of thalassemia with anemia rather than erythrocytosis.

In Hb Chesapeake, identified by Charache and co-workers (1) in an elderly man (and in other members of the family) with erythrocytosis, leucine replaces arginine at α92. The functional properties of Hb Chesapeake include 1) increased oxygen affinity in isolated Hb Chesapeake, unfractionated hemolysates and whole red cells, 2) well preserved Bohr effect (decrease in oxygen affinity with decrease in pH) and 3) decreased cooperative interactions with value of n from Hill's equation of 1.3.*

* Cooperative interactions, or heme-heme interaction, refers to the interactions in the Hb molecule which account for the sigmoid shape of the O_2 dissociation curve. The rearrangements of the polypeptide chains during reactions of Hb with oxygen rather

Perutz has recently reviewed his X-ray crystallographic evidence of the structural differences between oxy- and deoxyhemoglobin (2, 3): the amino acid substitution in Hb Chesapeake occurs in the area of contact between the α_1- and β_2-polypeptide chains. The area of α_1-β_2 contact undergoes considerable rearrangement during the reactions of hemoglobin with oxygen, and the substitution appears to interfere with the sliding of the polypeptide chains past one another during these reactions. From studies of the kinetics of the reactions of Hb Chesapeake with ligands, the suggestion has been made that its conformation might be intermediate between the oxy and deoxy forms of Hb A (4).

Other hemoglobin variants in which increased oxygen affinity has been associated with erythrocytosis are listed in Table 1 (5–8). Some of the affected patients had received radioactive phosphorous for their erythrocytosis. With the exception of Hb Hiroshima, the hemoglobins exhibiting high oxygen affinity have shown fairly good preservation of the Bohr effect and decreased heme-heme interaction. The amino acid substitutions in stable variants with high oxygen affinity have occurred in 1) the contact between the α_1-and β_2-chains or 2) the residues in the final (H) helix of the β-chain.

One hemoglobin variant has been shown to be associated with low oxygen affinity and cyanosis, Hb Kansas (10, 11) in which, because of decreased oxygen affinity, the hemoglobin is not fully saturated with oxygen at the partial pressures of oxygen found in the atmosphere. Whereas decreased dissociation of Hb Chesapeake to dimers has been demonstrated (12), such deaggregation occurs more readily in Hb Kansas than in normal Hb (11). Although the role of dimers in the reactions of hemoglobin with oxygen under physiological conditions is uncertain, the extent of deaggregation provides information about differences in conformation of high and low affinity hemoglobins.

During the last two years, the studies of Chanutin and Curnish (13) and of Benesch and Benesch (14, 15) have indicated that the levels of organic polyphosphates of the red cell, particularly 2, 3-diphosphoglycerate (DPG) are important in regulating the oxygen affinity of normal hemoglobin. Deoxyhemoglobin appears to bind DPG, and increased red cell levels of

than the direct effect of one heme upon another appear to be responsible for this change in oxygen affinity with oxygenation of Hb. The value of n from Hill's equation may be taken as an overall empiric measurement of cooperative interactions. The maximum value for n in a tetrameric Hb would be 4.0, and in the absence of interactions n would be 1.0. Values for n of about 3.0 are obtained in studies of normal human Hb.

TABLE 1. Hb associated with erythrocytosis

Hb	Amino acid substitution					Bohr effect	Cooperative interactions (n values)	O_2 affinity	Ref.
	Chain	Residue	Helical[a] residue	Normal amino acid	Substituted amino acid				
Chesapeake	α	92	FG4	Arg	Leu	Normal	1.3	Increased	(1)
Yakima	β	99	G1	Asp	His	Near normal	1.0	Increased	(5)
Kempsey	β	99	G1	Asp	Asn	Near normal	1.1	Increased	(6)
Hiroshima	β	143	H21	His	Asp	Reduced	2.3	Increased	(7)
Ranier	β	145	H23	Tyr	His	Normal	1.2	Increased	(8)

[a] In the notation of Perutz (9), the helical areas of the chains are designated by letters beginning at the N-terminal and the interhelical areas carry the letters of the two adjacent helical areas. Each residue is then numbered within the helix, e.g. E7 is the seventh (from the N-terminal) residue of the E helix. This system preserves the homology among chains: F8 is the proximal heme-linked histidine of both α- and β-chains.

TABLE 2. Some properties of the M hemoglobins

Hb M	Substitution				O_2 reactive chain	O_2 affinity at $P_{\frac{1}{2}}$	Bohr effect	Ref. for O_2 equilibria	50% oxidation of Hb CO M by ferricyanide compared with CO Hb A [a]
	Amino acid	Chain	Helical residue						
Boston	Tyr	α	E7		β	Decreased	Absent	(16)	Very fast
Iwate	Tyr	α	F8		β	Decreased	Absent	(17)	Not done
Saskatoon	Tyr	β	E7		α	Near normal	Present	(18)	Fast
Hyde Park	Tyr	β	F8		α	Near normal	Present	(19)	Fast
Milwaukee-1	Glu	β	E11		α	Decreased	Present	(20)	Very fast

[a] Ref. (20, 21)

DPG are associated with a decrease in the O_2 affinity of the hemoglobin. Decreased oxygen affinity facilitates the delivery of oxygen to the tissues. Thus far DPG has had the expected effect of lowering oxygen affinity in the few hemoglobin variants in which the effect has been tested.

M HEMOGLOBINS

The M hemoglobins, in which one pair of hemes (either both α or both β) are in the oxidized form because of an amino acid substitution in the polypeptide chains, are unusual hemoglobins for observations of structure-function relationships. Of the five M hemoglobins which have thus far been identified, four result from the substitution of tyrosine for the proximal (F8) or distal (E7) heme-linked histidine residues. A fifth, Hb $M_{Milwaukee-1}$ (Hb M_{M-1}) results from the substitution of glutamic acid for valine at residue E11 of the β-polypeptide chain. As a result of the amino acid substitution, the heme of the affected chain is in the oxidized form and hence unreactive toward oxygen, and the equilibria with oxygen are properties of the normal chains in the tetramer. Qualitative data concerning some properties of the M hemoglobins are given in Table 2.

The M hemoglobins substituted at β F8 or β E7, in which only the α chains undergo reactions with oxygen, have a nearly normal oxygen affinity at $P_{\frac{1}{2}}$ (the partial pressure of oxygen at which half the hemoglobin in the solution is oxygenated) with decreased cooperative interactions and a well preserved Bohr effect. (Values of n from 1.1 to 1.3 have been observed but since only two hemes in the tetramer can react with oxygen, the meaning of the low n values is uncertain). The α-substituted M hemoglobins exhibit absence of the Bohr effect in the physiologic range of pH together with low oxygen affinity (at $P_{\frac{1}{2}}$) and decreased cooperative interactions. Hb M_{M-1}, unlike the other β-substituted M hemoglobins, exhibits a low O_2 affinity, although it has a similar Bohr effect.

The data obtained from studies of oxygen equilibria of the M hemoglobins suggest that the α- and β-chains might differ in their oxygenation function in the normal hemoglobin tetramer. However, the oxygen affinities of the β-substituted M hemoglobins are themselves different: the affinities of the M hemoglobins carrying tyrosine substitution for either heme-linked histidine of the β chain resembles Hb A at $P_{\frac{1}{2}}$, while Hb M_{M-1} has a low O_2 affinity. It is possible that the oxidation of α-hemes induces different restrictions on the β-chain partners than does oxidation of the β-chains. Included in Table 2 are qualitative descriptions of the rate of reaction of several of the CO Hb M derivatives with ferricyanide. This reagent would oxidize the normal (CO) subunits in each of the M hemoglobins, and in each case the rapid reaction suggests some conformational differences of the normal subunit from Hb A. The applicability of the data on the M hemoglobins to considerations of the equivalency of the α- and β-chain of normal Hb in O_2 reactivity is open to question on the basis of this and other reactions of the normal subunits (20, 21).

The recent studies of Perutz (22) indicate that about half of the normal Bohr effect may be explained by altered ionization of the C-terminal histidines of the β-chains which are free in oxy-hemoglobin and bonded to β94 aspartates (of their own chains) in deoxyhemoglobin. The absence of a Bohr effect in the α-substituted M hemoglobins suggests that the oxygenation-dependent changes in position of the β C-terminal histidines do not occur in these M hemoglobins. The well preserved Bohr effect found in the M hemoglobins which contain oxygen reactive α-polypeptide chains suggests that the oxygenation-dependent conformational changes in the β146 histidine residues may be initiated by reactions of the α-chains with oxygen, even though the β-hemes are in the oxidized form.

The author thanks Dr. R. L. Nagel, Mrs. L. Udem, Mrs. S. Kumin, Dr. P. Heller and Dr. A. Pisciotta who collaborated in the studies carried out in the author's laboratory.

Supported by Grants AM 04502 and 1 PO2 AM13430 from the National Institutes of Arthritis and Metabolic Diseases and by Grant G–6936 from the Life Insurance Medical Research Fund.

REFERENCES

1. CHARACHE, S., WEATHERALL, D. J. and CLEGG, J. B. Polycythemia associated with a hemoglobinopathy. *J. clin. Invest.* **45**: 813, 1966.
2. PERUTZ, M. F. The Croonian Lecture, 1968. The haemoglobin molecule. *Proc. roy. Soc. B*, **173**: 113, 1969.
3. PERUTZ, M. F. Structure and function of hemoglobin. *Harvey Lect.* **63**: 213, 1969.
4. NAGEL, R. L., GIBSON, Q. H. and CHARACHE, S. Relation between structure and function in hemoglobin Chesapeake. *Biochemistry* **6**: 2395, 1967.
5. JONES, R. T., OSGOOD, E. E., BRIMHALL, B. and KOLER, R. D. Hemoglobin Yakima. I. Clinical and biochemical studies. *J. clin. Invest.* **46**: 1840, 1967.
6. REED, C. S., HAMPSON, R., GORDON, S., JONES, R. T., NOVY, M. J., BRIMHALL, B., EDWARDS, M. J. and KOLER, R. D. Erythrocytosis secondary to increased oxygen affinity of a mutant hemoglobin, hemoglobin Kempsey. *Blood* **31**: 623, 1968.
7. HAMILTON, H. B., IUCHI, I., MIYAJI, T. and SHIBATA, S. Hemoglobin Hiroshima (β143 histidine → aspartic acid): A newly identified fast moving beta chain variant associated with increased oxygen affinity and compensatory erythremia. *J. clin. Invest.* **48**: 525, 1969.
8. STAMATOYANNOPOULOS, G., YOSHIDA, A., ADAMSON, J. and HEINENBERG, S. Hemoglobin Ranier (β145 tyrosine → histidine): Alkali-resistant hemoglobin with increased oxygen affinity. *Science* **159**: 741, 1968.

9. PERUTZ, M. F.   Structure and function of haemoglobin. I. A tentative atomic model of horse oxyhaemoglobin. *J. molec. Biol.* **13**: 646, 1965.
10. REISSMANN, K. R., RUTH, W. E. and NOMURA, T.   A human hemoglobin with lowered oxygen affinity and impaired heme-heme interactions. *J. clin. Invest.* **40**: 1826, 1961.
11. BONAVENTURA, J. and RIGGS, A.   Hemoglobin Kansas, a human hemoglobin with a neutral amino acid substitution and an abnormal oxygen equilibrium. *J. biol. Chem.* **243**: 980, 1968.
12. BUNN, H. F.   Subunit dissociation of certain abnormal human hemoglobins. *J. clin. Invest.* **48**: 126, 1969.
13. CHANUTIN, A. and CURNISH, R. R.   Effect of organic and inorganic phosphates on the oxygen equilibrium of human erythrocytes. *Arch. Biochem.* **121**: 96, 1967.
14. BENESCH, R. and BENESCH, R. E.   Effect of organic phosphates from human erythrocytes on allosteric properties of hemoglobin. *Biochem. biophys. Res. Commun.* **26**: 162, 1967.
15. BENESCH, R., BENESCH, R. E. and YU, CHI ING.   Reciprocal binding of oxygen and diphosphoglycerate by human hemoglobin. *Proc. nat. Acad. Sci. (Wash.)* **59**: 526, 1968.
16. SUZUKI, T., HAYASHI, A., YAMAMURA, Y., ENOKI, Y. and TYUMA, I.   Functional abnormality of hemoglobin M_{Osaka}. *Biochem. biophys. Res. Commun.* **19**: 691, 1965.
17. HAYASHI, N., MOTOKAWA, Y. and KIKUCHI, G.   Studies on relationships between structure and function of hemoglobin M_{Iwate}. *J. biol. Chem.* **241**: 79, 1966.
18. SUZUKI, T., HAYASHI, A., SHIMIZU, A. and YAMAMURA, Y.   The oxygen equilibrium of hemoglobin $M_{Saskatoon}$. *Biochim. biophys. Acta (Amst.)* **127**: 280, 1966.
19. RANNEY, H. M., NAGEL, R. L., HELLER, P. and UDEM, L.   Oxygen equilibrium of hemoglobin $M_{Hyde\ Park}$. *Biochim. biophys. Acta (Amst.)* **160**: 112, 1968.
20. UDEM, L., RANNEY, H. M., BUNN, F. H. and PISCIOTTA, A.   Some observations on the properties of haemoglobin $M_{Milwaukee-1}$. *J. molec. Biol.* **48**: 489, 1970.
21. HAYASHI, A., SUZUKI, T., SHIMIZU, A. and YAMAMURA, Y.   Properties of hemoglobin M. Unequivalent nature of the α and β subunits in the hemoglobin molecule. *Biochim. biophys. Acta (Amst.)* **168**: 262, 1968.
22. PERUTZ, M. F., MUIRHEAD, H., MAZZARELLA, L., CROWTHER, R. A., GREER, J. and KILMARTIN, J. V.   Identification of residues responsible for the alkaline Bohr effect in haemoglobin. *Nature (Lond.)* **222**: 1240, 1969.

HEMOGLOBIN TORINO DISEASE
Clinical and Biochemical Findings

E. GALLO, G. RICCO, V. PRATO, G. BIANCO and U. MAZZA

Institute of Medical Pathology, University of Turin, Turin, Italy

During the last few years, several cases of congenital nonspherocytic hemolytic anemia due to unstable hemoglobins have been described. In 1968, we reported Hb Torino which is a labile hemoglobin with a substitution in its α-chain of phenylalanine by valine at position CD 1 (α 43) (1).

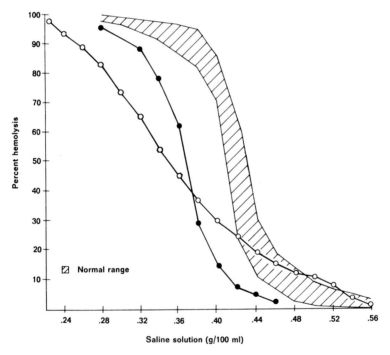

FIG. 1. Red cell osmotic fragility of the propositus. —•— before splenectomy, —o— after splenectomy.

The clinical, hematological and biochemical findings in carriers of this abnormality are summarized in this paper.

The propositus, a 36-year-old man, was first admitted in 1959 when he was found to have a hemolytic anemia. The patient was poorly nourished, pale and subicteric. Some features of mongoloid facies were apparent. The spleen was palpable 4 cm below the costal margin and the liver was just palpable. The hematological examination revealed a Hb of 11.2 g/100 ml; RBC 3,560,000/mm^3, packed cell volume (PCV) 33%; mean corpuscular Hb 31.4 pg; mean corpuscular Hb concentration 34%; mean corpuscular volume 92 µ^3; reticulocytes 6.5%. Total bilirubin was 1.3 mg/100 ml; serum iron 210 µg/100 ml; total iron binding capacity (TIBC) 468 µg/100 ml; methemoglobin 4.6%; methemoglobin after 48 hr incubation of the blood at 37 C was 48.0%. (Our normal values for methemoglobin are 1.75% ± 0.78 in fresh samples and 6.96% ± 3.40 after 48 hr incubation.) The T$^1/_2$ of the patient's RBC tagged with Cr^{51} was 12 days. The RBC showed an increased osmotic resistance (Fig. 1). Bone marrow biopsy revealed a hyperplasia of the erythroid series, the majority of the cells being polychromatic and orthochromatic normoblasts.

X-ray of the skull showed a thickening of the diploe and trabecular formations similar to the "brush cranium" found in Cooley's anemia.

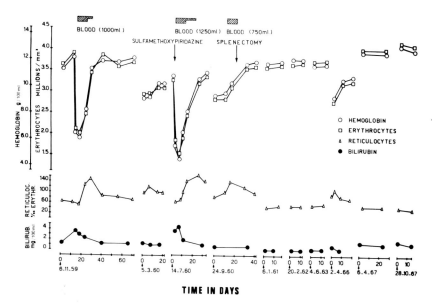

FIG. 2. Clinical course of the propositus before and after splenectomy.

The clinical course of the disease, followed over eight years, is summarized in Fig. 2. The patient was first hospitalized during a sudden and severe hemolytic crisis, accompanied by epigastric pain, nausea and fever. The hemoglobin level fell to 6 g/100 ml, and the patient was treated with blood transfusions. A reticulocyte response as well as a release of polychromatic and orthochromatic normoblasts into the peripheral blood was also observed. During the following six months, two similar episodes were observed, one of which (July 1960) followed the administration of 1.5 g sulphamethoxy piridazine. On this occasion about 60% of the peripheral RBC showed Heinz bodies. Examinations performed several times after this crisis showed that these inclusion bodies were no longer present. In October 1960, after another crisis, splenectomy was performed. The operation was well tolerated and since then the patient has been well and has worked for seven years. Except during one very mild hemolytic crisis he has not required any blood transfusions. Hemoglobin level was steadily over 12 g/100 ml and the Cr^{51} red cell survival was 16 days instead of 12 days. After splenectomy, Heinz bodies were constantly present in about 60% of the RBC.

RBC osmotic fragility behaved differently before and after splenectomy: in fact, while before splenectomy the whole hemolytic curve was shifted to the left, showing an increased resistance of RBC, after splenectomy the curve shifted towards both higher and lower tonicity (Fig. 1).

In addition to the propositus, two other members of the family were affected: the mother and a sister.

The disease was inherited from the mother, a woman aged 69, who has been pale all her life. Her hematological data were as follows: Hb 9.5 g/100 ml; PCV 27%; reticulocytes 4.5%. The spleen was just palpable. She had been pregnant five times and never had any abortions. On the other hand, the sister of the propositus had a long history of hemolytic crises associated with excretion of dark urine and had been admitted to the hospital several times. In 1954, splenomegaly was observed: her Hb level was 7.7 g/100 ml; PCV 24% and reticulocytes 14%. Recently she was re-examined and her Hb was 9.0 g/100 ml, PCV 30% and reticulocytes 12%. The spleen was palpable 5 cm below the costal margin. In spite of the anemia she had been pregnant three times and never had any abortions.

The features of the anemia in these two cases were similar to those described in the propositus: the methemoglobin level was increased in the fresh blood and the level increased considerably on incubation for 48 hr. RBC osmotic resistance was increased and the heat denaturation test was positive. We were unable to demonstrate Heinz bodies in the RBC of these

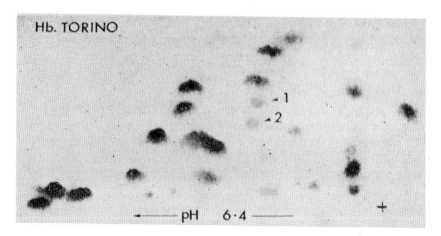

FIG. 3. Fingerprint of partially purified Hb Torino (1) α^A Tp VI; (2) α^{Torino} Tp VI.

two subjects, even during the hemolytic crises. When their blood was incubated with acetylphenyl-hydrazine, more Heinz bodies appeared as compared to normal RBC incubated under identical conditions. This was particularly evident in the sister of the propositus.

Hb studies. Hemoglobin electrophoresis on starch gel did not reveal any apparent abnormality. By the heat denaturation test, according to the method of Dacie (2), about 20% of the hemoglobin precipitated in insoluble form. The presence of Heinz bodies, the increased amounts of methemoglobin and the tendency to form methemoglobin on incubation, together with the fact that the hemoglobin was precipitated by mild heating, suggested that the cause of the anemia was an unstable hemoglobin. Further analysis showed that only 8% of the precipitate belonged to the abnormal hemoglobin.

The heat unstable material was fingerprinted (Fig. 3) and the map chromatogram showed a new peptide situated below the normal α VI which contains the residues 41 to 56. Amino acid analysis demonstrated that a phenylalanine was replaced by a valine. As α VI contains two phenylalanines, in position 43 and 46, it was necessary to verify the exact location of the phe $\rightarrow$ val substitution. The abnormal peptide was treated by the combined Dansyl-Edman sequencing method (3) and the mutation was shown to be present in residue 43 (CD1).

Phenylalanine in position CD 1 of the α-chain lies close to the porphyrin of the heme and is one of the amino acids which form the heme pocket and protect the heme from oxidation. A replacement in this position by a valine, as suggested by Perutz and Lehmann (4), would disturb the conformation of the CD segment making the whole subunit unstable. This explains the instability of the molecule when heated at 50 C. Moreover, the disarrangement of the heme pocket makes the molecule more easily oxidized to methemoglobin, either by mild oxidizing agents such as sulphonamides (which are able to cause a hemolytic crisis with spontaneous occurrence of Heinz bodies), or spontaneously, as demonstrated by the rise of methemoglobin level in samples incubated in sterile tube for 48 hr at 37 C.

Hb Torino behaved, on electrophoresis, in the same way as Hb A, only when the hemolysate was tested immediately after blood collection or when potassium cyanide was added. On the other hand, when electrophoresis was performed two or three days after blood collection, a new fraction moving close to Hb A was observed. In one sample only, namely that of the sister of the propositus, we observed a second, slow moving fraction amounting to about 1% of the total hemoglobin which migrates a little slower than Hb S.

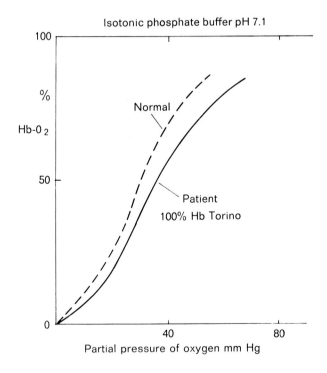

FIG. 4. Oxygen dissociation of red cell from a patient with Hb Torino.

While the fraction moving close to Hb A can be considered as methemoglobin, the nature of the fraction moving in S position is not completely clear. Schneider et al. recently described the unstable hemoglobin Sabine (5); in this case the mobility of the two abnormal fractions are considered to be due to their deficiency in heme content. Even if we do not have any experimental evidence, it seems reasonable to believe that in our case the presence of the minute fraction has a similar explanation.

Studies were carried out on the oxygen dissociation properties of these RBC. (The study was performed by Drs. Huehns and Bellingham at the University College Hospital Medical School in London). It was found that Hb Torino has a low oxygen affinity (Fig. 4). According to Huehns and Bellingham (6) oxygen affinity in the RBC is one of the factors which determine the degree of compensation in hemolytic anemia in the steady state. From the point of view of oxygen delivery, the RBC of our patient would release to the tissue more oxygen per gram of hemoglobin than normal RBC, thus compensating for the low PCV.

Porphyrin synthesis and metabolism. The results of these studies are summarized in Tables 1 and 2. The free RBC porphyrin level was greatly increased, while δ-amino levulinic acid (ALA) and porphobilinogen (PBG) excretion was within normal limits. The urinary and fecal excretion of coproporphyrin and the fecal excretion of protoporphyrin were lower than normal.

In the synthesis experiment we found that total PBG synthesis (PBG + porphyrin) obtained by incubation of the RBC *in vacuo* with a nonlimiting amount of ALA was within normal limits, and that the synthesized PBG was converted to porphyrins faster than normal; the total PBG obtained in systems with limiting amount of ALA is $1/3$ of the normal amount, while the synthesis of porphyrin from PBG is not decreased.

SUMMARY AND CONCLUSIONS

The clinical and hematological features of the disease are similar to those described in other cases of nonspherocytic hemolytic anemia due to various unstable hemoglobins. However, we would like to mention some aspects which characterize the described family. Although the propositus had several hemolytic crises, only in one instance were we able to recognize the determining cause; i.e. the sulphonamide administration. On that occasion about 60% of erythrocytes presented Heinz bodies, which quickly disappeared from the peripheral blood and never appeared again until the

TABLE 1. *Free erythrocyte porphyrins, urinary δ-aminolaevulic acid (ALA), porphobilinogen (PBG) and coproporphyrin and fecal copro- and protoporphyrin of propositus after splenectomy*

	Erythrocytes		Urine			Stools	
	Coproporphyrin	Protoporphyrin	ALA	PBG	Coproporphyrin	Coproporphyrin	Protoporphyrin
	μg/100 ml			μg/24 hr		μg/g dry feces	
Normal values: mean	1.25	36.22	2,844	671	102.40	7.54	14.48
SD	±1.00	±15.36	±1,228	±414	±29.30	±3.46	±5.29
Propositus	2.48	213.52	2,418	741	36.00	3.60	5.60

TABLE 2.

A) *Synthesis in vitro of PBG and porphyrins in erythrocytes incubated in vacuo at 37 C for 60 min with nonlimiting amounts of ALA*

	ALA in system		Erythrocytes, mg/100 ml		
	before incubation	after incubation	PBG synthesized	Porphyrin synthesized	Porphyrin + PBG
Normal values: mean	in excess	not detected	9.74	1.32	11.06
range			(9.13 to 10.68)	(0.99 to 1.64)	—
Propositus	in excess	not detected	4.80	4.50	9.30

B) *Synthesis in vitro of PBG and porphyrins in erythrocytes incubated in vacuo at 37 C for 60 min with limiting amounts of ALA*

	ALA in system		μg in system		
	before incubation	after incubation	PBG synthesized	Porphyrin synthesized	Porphyrin + PBG
Normal values: mean	115	77.07	30.0	3.62	34.22
range		(71.0 to 83.0)	(27.0 to 35.2)	(3.20 to 4.42)	—
Propositus	115	70.14	7.22	4.65	11.87

patient was splenectomized. Since then, Heinz bodies became a constant feature of the disease. Hb Torino patients are similar to Hb Zürich patients in their hemolytic response to mild oxidizing agents such as sulphonamide (7). Heinz bodies, however, were absent during all the other hemolytic crises observed in the propositus or affected relatives.

In spite of the low level of hemoglobin, the anemia was well tolerated by the patients. For instance the mother of the propositus never complained of severe hemolytic crises and was pregnant five times. The propositus himself did not complain of any important disorders before he was 20 years old and after splenectomy, was able to work, never needing blood transfusions. The sister, in spite of several hemolytic crises, had three pregnancies with no abortions.

Several factors are probably the cause for the mildness of the disease: the phenylalanine → valine mutation (both neutral and nonpolar residues); the fact that the α-chain heme is less easily removed than the β-chain heme (8) and finally the normal β-chain may extend some protection even to unstable α-chains during the stepwise oxidation of the hemoglobin. The latter occurs first on the β 93 cysteine, subsequently on the β 112 cysteine, and later on, on the α 104 cysteine. The small amount of unstable hemoglobin (8 to 10%) and finally the low oxygen affinity of the RBC, which are able to release more oxygen per gram of hemoglobin than normal RBC, are probably the cause of the mild clinical syndrome in these patients. The disturbance in the heme synthesis and porphyrin metabolism are probably related to the globin abnormality.

REFERENCES

1. BERETTA, A., PRATO, V., GALLO, E. and LEHMANN, H. Haemoglobin Torino — α 43 (CD 1) phenylalanine → valine. Nature (Lond.) **217**: 1016, 1968.
2. DACIE, J. V., GRIMES, A. J., MEISLER, A., STEINGOLD, L., HEMSTED, E. H., BEAVEN, G.H. and WHITE, J. C. Hereditary Heinz-bodies anaemia: a report of studies on five patients with mild anaemia. Brit. J. Haematol. **10**: 388, 1964.
3. GRAY, W. R. and HARTLEY, B. S. The structure of chymotryptic peptides from Pseudomonas cytochrome c55. Biochem. J. **89**: 379, 1963.
4. PERUTZ, M. F. and LEHMANN, H. Molecular pathology of human haemoglobin, Nature (Lond.) **219**: 902, 1968.
5. SCHNEIDER, R. G., MUEDA, S., ALPERIN, J. B., BRIMHALL, B. and JONES, R. T. Haemoglobin Sabine: beta 91 (F 7) Leu → Pro. An unstable variant causing severe anemia with inclusion bodies. New Engl. J. Med. **280**: 739, 1969.
6. HUEHNS, E. R. and BELLINGHAM, A. J. Disease of function and stability of haemoglobin. Brit. J. Haematol. **17**: 1, 1969.
7. FRICK, P. G., HITZIG, W. H. and BETKE, K. Haemoglobin Zürich. A new hemoglobin anomaly associated with acute hemolytic episodes after sulphonamide therapy. Blood **20**: 261, 1962.
8. CASSOLY, R., BUCCI, E., IWATSUBO, M. and BANERJEE, R. Functional studies on human semi-hemoglobin. Biochim. biophys. Acta (Amst.) **133**: 557, 1967.

PROPERTIES OF OXIDIZED HEMOGLOBIN SUBUNITS SEPARATED *IN VITRO* AND THEIR RELATION TO INTRAERYTHROCYTIC INCLUSION BODIES IN THALASSEMIA

ELIEZER A. RACHMILEWITZ

Departments of Hematology and Medicine, Hadassah University Hospital and Hebrew University-Hadassah Medical School, Jerusalem, Israel

Tetrameric human hemoglobin consists of four subunits, each bearing a heme group. The interactions among these subunits, and between the subunits and the heme groups are important in the production of the three critical functional properties of hemoglobin: heme-heme interaction, the Bohr effect and low oxygen affinity. These interactions are probably also important in the synthesis of hemoglobin and in this respect have a bearing on several hemolytic syndromes associated with abnormal hemoglobin synthesis and structure.

The following examples emphasize the importance of interchain contacts between unlike polypeptide chains: 1) Hb H, a tetramer which consists only of β-chains, lacks all three critical functions mentioned above (1). 2) Mammalian hemoglobins generally have lower oxygen affinity than many vertebrate hemoglobins and mammalian myoglobins. However, in Hb H as well as in monomeric forms of α- and β-chains of Hb A, high oxygen affinity is present (1,2). 3) Perutz and Lehmann (3) have indicated that the amino acid sequence in some of the unstable hemoglobin variants may affect not only the binding of heme to globin and the integrity of the helical structure but also the interchain contacts. The unstable Hb Philly (β 35 tyrosine → phenylalanine), in which there is a loss of one normal hydrogen bond at the α_1-β_1 contact, is a good example of this latter concept (4).

Our interest in the importance of interchain interactions stemmed from the observation that the oxidation of isolated hemoglobin subunits follows a different pattern from the oxidation of a tetramer of Hb A. The results and implications of our studies may be summarized as follows:

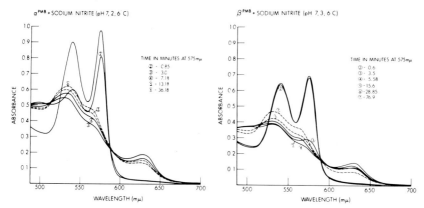

FIG. 1. Spectra of α– (right) and β– (left) PMB chains after the addition of sodium nitrite (1:1 M ratio/heme). Note the appearance of hemichrome spectra (interrupted lines) 36 and 77 min after addition of oxidizing agent to the α– and β–PMB chains, respectively.

a) Isolated subunits of Hb A, still bearing heme groups, did not form stable ferrihemoglobin when oxidized with an appropriate oxidizing agent such as ferricyanide or sodium nitrite. Instead, absorption spectra characteristic of heme proteins, known as hemichromes, were seen after the transient appearance of ferrihemoglobin (Fig. 1). This difference in behavior between Hb A and isolated hemoglobin subunits suggested that the normal tetrameric structure of Hb A is important in preventing hemichrome formation (5). Hemichromes are heme proteins characterized by a main absorption peak around 530 to 535 mμ and a shoulder near 565 mμ in the optical region of the spectrum. Hemichrome formation in Hb A is thought to be associated with the binding of the heme iron to a histidyl residue (the opposing histidine E7 in the amino acid sequence). The binding becomes possible due to a small change in the tertiary structure of the subunit since histidine E7 is not ordinarily in contact with the heme iron, (6) (Fig. 2). Since tetramerization in Hb A prevents hemichrome formation, it would appear that the quaternary structure (interaction between subunits) is related to the structure of individual subunits: association of isolated subunits prevents the change in the tertiary structure leading to the formation of hemichromes (5).

b) Three heme-bearing hemoglobin subunits, which are the main components of adult and fetal hemoglobins, have been oxidized and in all of them hemichromes were formed, as demonstrated by optical and electron spin resonance (ESR) spectroscopy (5,7). However, the rate of hemichrome formation and the stability of hemichromes in solution were not the same

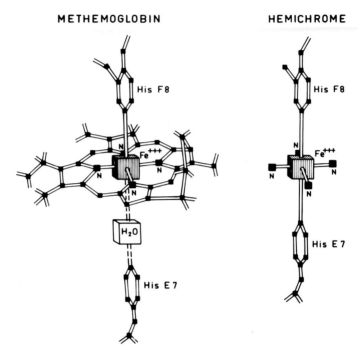

FIG. 2. A part of a hypothetical model illustrating the possible change in the structure of oxidized hemoglobin subunit around the heme group, resulting in removal of the water molecule and leading to the formation of hemichrome.

His F8 and E7 represent the two histidines in relation to the heme iron in methemoglobin and in hemichrome. The double iron-histidine bond is generally thought to be the structure of heme-proteins which have the absorption spectrum of a hemichrome.

for each subunit (Fig. 1,3). γ-chains appeared to be most stable in the form of ferrihemoglobin and it was only after 24 hr that slight changes in the ferrihemoglobin spectrum were observed: a decrease in the two main absorption peaks of ferrihemoglobin at 630 mμ and 500 mμ and an increase between 565 and 535 mμ. On the other hand, both oxidized α- and β-chains formed significant amounts of hemichrome within minutes at room temperature and within half an hour to an hour at 4 C (Fig. 1,3). This latter reaction occurred irrespective of whether the method of preparing the subunits was by recombining heme to isolated α- or β-globin chains or by separating heme-bearing subunits with paramercuribenzoate (PMB) (5). The formation of hemichromes was even better demonstrated by the use of ESR spectroscopy. With this technique, ferrihemoglobin A produced a high spin signal

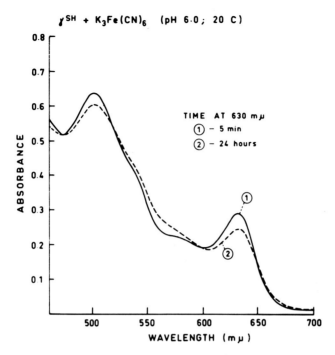

FIG. 3. Spectra of γSH-chains (prepared by Dr. R. Banerjee, Paris) separated with PMB from cord blood hemoglobin (8). The first spectrum (straight line) was recorded 5 min after the addition of ferricyanide to oxygenated γ-chains. The second spectrum (dotted line) was recorded 24 hr later. Note the decrease in 630 and 500 mµ absorption peaks and the increase between 560 to 530 mµ in the second spectrum.

around $g = 6$ whereas hemichrome, in which both fifth and sixth coordination positions of heme iron are bound to ligands, gave a low spin signal around $g = 2$. The spectrum of the two oxidized compounds did not overlap as in the optical region of the spectrum. Another advantage of ESR spectroscopy is that it does not record any traces of oxyhemoglobin present in the solution, while even small amounts of oxyhemoglobin will affect the shape of the curve in the visual area of the spectrum, mainly around 576 mµ and 540 mµ. While α- and β-subunits, with and without PMB used for their separation, were unstable and tended to precipitate shortly after the hemichromes began to form, γ-subunits remained in a soluble form for days and in this respect resembled ferrihemoglobin A (Fig. 3).

c) Certain differences occurred between oxidation of α-PMB and β-PMB chains. α-chains formed hemichromes faster than β-chains. This may be a

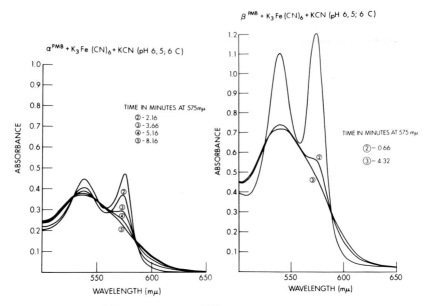

FIG. 4. Spectra of α^{PMB}-chain (4a) and β^{PMB}-chain (4b) after the simultaneous addition of ferricyanide and cyanide. Note that cyanmethemoglobin was formed from each subunit after different time intervals.

consequence of the greater tendency of β-chains than α-chains to aggregate.

d) Ferrihemoglobin cyanide was formed from hemichromes with the addition of cyanide. In addition, when ferricyanide and cyanide were simultaneously added in 4:1 M excess per heme to oxygenated α- or β-hemoglobin subunits, stable cyanmethemoglobin subunits were formed and no hemichromes were detected (Fig. 4). These results indicate that the cyanide ion can displace the histidyl side chain from the sixth position of the heme iron and not vice versa. These observations are in accordance with data in the literature on the affinity of the sixth coordination position of heme iron for various ligands (9).

e) The absorption spectra of the hemichromes did not change with changes in pH between 6 and 9, a finding which is consistent with the unavailability of the sixth position of the heme iron. The spectrum of ferrihemoglobin A is known to change with pH. This is associated with the removal of a proton attached to the water molecule normally bound to the sixth position. This supports the identification of our derivatives of hemoglobin as hemichromes.

f) Differences in the ultra violet spectra of proteins reflect changes in

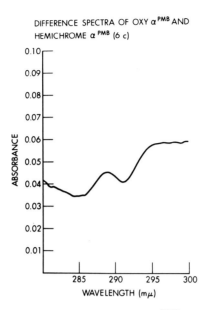

FIG. 5. Ultraviolet difference spectra of oxygenated α^{PMB} and the hemichrome of α^{PMB}. Note an absorption peak at 291 mμ, characteristic of perturbation of a tryptophan residue and reflecting changes in structure.

their structure. In particular, perturbation of tyrosin and tryptophan exists in Hb A as a function of changes in pH and state of oxygenation (10, 11). Difference spectra of hemichromes (as compared with other derivatives in known conformational states) can reveal aspects of structure. A difference spectrum with a peak at 291 mμ, characteristic of a perturbation of tryptophan residue, was found between hemichromes and oxygenated ferrihemoglobin subunits, indicating a change in interchain relations and interactions (Fig. 5). However, this type of change is not specific for hemichromes only; as it is similar to the difference spectra seen with deoxygenation of Hb A (10).

In addition to the importance of the properties of oxidized hemoglobin subunits in influencing structure-function relationships in the hemoglobin molecule, the question arises as to whether these observations have any bearing on the molecular pathology of hemoglobins associated with various hemolytic disorders. A logical starting point for such an investigation is the thalassemia syndrome, in which the presence of an excess of one hemoglobin subunit inside a red cell is the result of the basic abnormality encountered in the syndrome. Thalassemic red cells may contain intraerythrocytic inclusion bodies, known to be breakdown products of hemoglobin (12). It has been shown that the inclusion bodies in β-thalassemia consist of specific peptides of α-chains (13), and that the inclusion bodies in a solution of red cell ghosts from patients with Hb H disease have the same electrophoretic mobility as β-globin chains (14). Following these observations our working hypothesis was that the inclusion bodies found in thalassemia were in effect hemichromes of either α-chains or β-chains. Optical and ESR spectroscopy of a solution of red cell ghosts containing numerous inclusion bodies confirmed unequivocally the presence of hemichromes in two patients with Hb H disease (α-thalassemia) (14). In another study, microspectrophotometry of single inclusion bodies in two cases of β-thalassemia major also revealed the characteristic spectrum of hemichrome (Rachmilewitz, E. A. and Thorell, B., in preparation). One can therefore conclude that there is a similarity between the properties of hemoglobin subunits prepared and oxidized *in vitro* and the behavior of the same subunits inside intact α- or β-thalassemic red blood cells.

To obtain sufficient amounts of inclusion bodies in α- or β-thalassemia for spectroscopic measurements, we had to incubate the red cells for 12 to 18 hr at 37 C. This procedure results in increased susceptibility of the cellular hemoglobin to oxidation and in this respect it can be compared to the addition of oxidizing agents to a hemoglobin solution *in vitro*. We do not as yet have sufficient data on the behavior of red cells containing an excess of Hb Barts (γ_4) under similar conditions and, as far as we are aware, inclusion bodies containing precipitated γ-chains have not been demonstrated. This may be due to the fact that oxidized heme-bearing γ-chains prepared *in vitro* were found to be more stable than either α- or β-chains.

The presence of inclusion bodies in thalassemic red blood cells is not only of academic interest but has a very important practical significance. It is believed that the inclusion bodies are bound to the red cell membranes by disulfide bonds and are later "pitted" in the sinusoids of the spleen (15). The removal of the inclusion bodies by the spleen results in damage to the red cell membrane. The presence of inclusion bodies in a thalassemic red blood cell is, therefore, important in the causation of the increased hemolytic tendency in this syndrome. Additional evidence for a relationship between the presence of inclusion bodies and the spleen is the fact that numerous inclusion bodies were readily demonstrated in a fresh preparation of red cell ghosts from a splenectomized patient with Hb H disease, whereas only a few inclusions could be seen in a similar preparation from an unsplenectomized patient. In order to demonstrate the inclusion bodies in the second case, the blood had to be incubated at 37 C for several hr (14, 16).

It is also known that concomitant with the presence of many inclusion bodies in splenectomized thalassemic patients, the red cell survival as measured with Cr^{51} was longer than in unsplenectomized patients (16). Another important observation was that of Nathan and Gunn (17) who showed that in β-thalassemic red blood cells the largest inclusion bodies were present in those cells which contained the least fetal hemoglobin. This suggests that when more γ-chains are present in a red cell some of the excess α-chains are bound to γ-chains and fetal hemoglobin is formed; there are fewer free α-chains and, therefore, fewer hemichromes, fewer and smaller inclusion bodies and a longer survival of this type of cell.

From all these data one can conclude that inclusion bodies in thalassemia play a major role in the causation of hemolytic anemia. The inclusion bodies are heme-bearing precipitated polypeptide chains in the form of hemichromes. The number and size of the inclusion bodies are dependent on the amount of free hemoglobin subunits within a red cell. This is controlled, on the one hand, by the degree of severity of the genetic defect leading to the impaired synthesis of either α- or β-chains and, on the other hand, by the ability to continue to synthesize γ-chains. In addition, the number and size of the inclusion bodies probably depend on the differences in the susceptibility of the various hemoglobin subunits to oxidation.

The author wishes to thank Dr. R. W. Briehl from Albert Einstein College of Medicine, Bronx, New York, for his help and suggestions and Dr. R. Banerjee from the Institut de Biologie Physico-Chimique, Paris, who kindly provided a sample of γ-chains.

Supported in part by a grant from the Henry and Lilian Stratton Research Foundation and J. S. Gordon Family Foundation, and by a special grant from the World Health Organization.

REFERENCES

1. BENESCH, R. E., RANNEY, H. M. and SMITH, G. M. The chemistry of the Bohr effect. II. Some properties of hemoglobin H. *J. biol. Chem.* **230**: 2926, 1961.
2. BRIEHL, R. W. The relation between oxygen equilibrium and aggregation in Lauprey hemoglobins (Abst.) *Fed. Proc.* **22**: 72, 1962.
3. PERUTZ, M. F. and LEHMANN, H. Molecular pathology of human hemoglobin. *Nature (Lond.)* **219**: 902, 1968.
4. RIEDER, R. F., OSKI, F. A. and CLEGG, J. B. Hemoglobin Philly (β 35 tyrosine-phenylalanine): Studies in the molecular pathology of hemoglobin. *J. clin. Invest.* **48**: 1627, 1969.

5. RACHMILEWITZ, E. A. Formation of hemichromes from oxidized hemoglobin subunits. *Ann. N.Y. Acad. Sci.* **165**: 171, 1969.
6. KENDREW, I. C. Three-dimensional structure of a protein. *Sci. Amer.* **205**: No. 6, 96, 1961.
7. RONISCH, P. and KLEIHAUER, E. Alpha-Thalassämie mit Hb.H und Hb. Bart's in einer deutschen Familie. *Klin. Wschr.* **45**: 1193, 1967.
8. KAJITA, A., KAZUTOSHI, T. and SHUKUYA, R. Isolation and properties of γ chain from human fetal hemoglobin. *Biochim. biophys. Acta (Amst.)* **175**: 41, 1969.
9. FALK, J. E. "Porphyrins and metalloporphyrins." Amsterdam, Elsevier Publishing Company, 1964, p. 48.
10. BRIEHL, R. W. and HOBBS, J. F. Ultraviolet difference spectra in human hemoglobins. I. Difference spectra in hemoglobin A and their relation to the function of hemoglobin. *J. biol. Chem.* **245**: 544, 1970.
11. BRIEHL, R. W. and RANNEY, H. M. Ultraviolet difference spectra in human hemoglobin. II. Difference spectra in isolated subunits of hemoglobin. *J. biol. Chem.* **245**: 555, 1970.
12. FESSAS, P., LOUKOPOULOS, D. and THORELL, B. Absorption spectra of inclusion bodies in thalassemia. *Blood* **25**: 105, 1965.
13. FESSAS, P., LOUKOPOULOS, D. and KALTSOYA, A. Peptide analysis of the inclusions of erythroid cells in beta thalassemia. *Biochim. biophys. Acta (Amst.)* **124**: 430, 1966.
14. RACHMILEWITZ, E. A., PEISACH, J., BRADLEY, T. B. and BLUMBERG, W. E. Role of haemichromes in the formation of inclusion bodies in haemoglobin H disease. *Nature (Lond.)* **222**: 248, 1969.
15. NECHELES, T. F. and ELLEN, D. M. Heinz body anemias. *New Engl. J. Med.* **280**: 203, 1969.
16. RIGAS, D. A. and KOLER, R. D. Decreased erythrocyte survival in hemoglobin H disease, as a result of the abnormal properties of hemoglobin H: The benefit of splenectomy. *Blood* **18**: 1, 1961.
17. NATHAN, D. G. and GUNN, R. B. Thalassemia: The consequences of unbalanced hemoglobin synthesis. *Amer. J. Med.* **41**: 815, 1966.

DISCUSSION

N. KOSOWER (*USA*): The paper presented by Dr. Rieder is open for discussion.

H. M. RANNEY (*USA*): Hemoglobin Philly, if it is unstable by virtue of increased dissociation to monomers, should show some evidence of increased dissociation without the addition of parachloromercurobenzoate (PMB). Do you have any data on its molecular weight?

R. F. RIEDER (*USA*): We have looked at the whole hemolysate for evidence of dissociation on Sephadex G-100 and in the ultracentrifuge. We have not been able to find any spreading of the peak in the ultracentrifuge. On Sephadex columns, in high or low salt solutions, we haven't been able to show any difference between a normal hemolysate and this hemolysate. Thus, there is no definite evidence of dissociation other than that caused by PMB.

H. M. RANNEY: Did you look at the oxy or deoxy in hemoglobin?

R. F. RIEDER: This was oxyhemoglobin. We have not attempted to examine deoxygenated hemoglobin and I must say, just in addition to this information, that it does appear that this hemoglobin has a high affinity for oxygen.

E. M. KOSOWER (*USA*): I would like to ask Dr. Rieder or anyone else who is willing to comment on it, how definite the evidence is that Heinz body formation within the cell really leads to the sequestration of the cell and not some other process, let us say, that is more closely associated with the membrane.

R. F. RIEDER: There is some electron micrographic evidence that seems to show that as the cells with Heinz bodies go through the spleen, they tend to be more rigid, more easily trapped in the spleen, and that macrophages tend to pull these Heinz bodies out of the red cells.

E. M. KOSOWER: Let us take any of the processes that lead to Heinz body formation; it seems to me that most of them involve some kind of oxidation. Whether it is the kind of mechanism that Dr. Rachmilewitz described, or another mechanism, there is some oxidation involved. Now what makes you think that this oxidation, whatever it is, is limited only to the inside of the cell? What you see inside the cell is the formation of Heinz bodies. But how do you know that associated with the reaction, there is not also a chemical challenge to the membrane which leads to the recognition of the cell by the spleen?

R. F. RIEDER: I think that since the mutation occurs in the hemoglobin molecule, one assumes that as a result these patients have hemolysis. One would assume from the known relationship of unstable hemoglobin to hemolysis that there is no primary genetic defect in the membrane in this disorder.

E. M. KOSOWER: What is known about glutathione content of cells which contain unstable hemoglobins or which produce Heinz bodies?

R. F. RIEDER: Patients with hemoglobin Zürich and hemoglobin Philly have been extensively studied for levels of many of the red cell enzymes and glutathione. The enzyme levels tend to be higher than normal and the amount of glutathione is higher than normal.

B. RAMOT (*Israel*): Exemplifying the importance of an excess of chains as a cause of damage to the red cell membrane are β-δ-thalassemia homozygotes. We had an opportunity to study three Arabs with homozygous β-δ-thalassemia. None of them was anemic and all had a compensated hemolytic process. I would like to answer Dr. Kosower about Heinz body formation and glutathione. In hemoglobin H disease and in one patient with an unstable hemoglobin that was not identified after splenectomy, the glutathione level in the heaviest ("oldest") cell was low. I think that during aging such cells lose more glutathione than do normal cells.

N. KOSOWER: Do β-δ-thalassemia patients have inclusion bodies?

B. RAMOT: They do not have any inclusion bodies on a 4-hr incubation.

E. KLEIHAUER (*West Germany*): I would like to make a short comment on reduced glutathione (GSH) content of red cells containing unstable hemoglobins. GSH has been reported to be low in Hb Köln disease while the activity of red cell glutathione reductase is increased. My question to Dr. Rieder is: The binding of heme to globin is very important for the stability of the globin molecule. In hemoglobin Gun Hill, the attachment is rather weak. I wonder why this is not correlated with a more pronounced and spontaneous Heinz body formation. Furthermore, is there any explanation for the finding that the red cell inclusions are so small and numerous? They look more like hemoglobin H inclusions than Heinz bodies typical for unstable hemoglobin types. The next question: Why is the abnormal β-chain synthesized to a higher proportion than the normal β-chain?

R. F. RIEDER: I really don't know why hemoglobin Gun Hill is as stable as it is. It may be that the α portion of the molecule, which is normal, stabilizes the tetramer via the α-β contact, so that precipitation of the whole molecule does not occur as rapidly as it would if the β-chains were separated from the α-chains. These β-chains without heme would precipitate very quickly. With regard to the size of the Heinz bodies formed, I can only say that this is variable and seems to depend to a great extent upon the source of the brilliant cresyl blue used. The reason for the more rapid synthesis of the β-chains of Gun Hill as compared to

DISCUSSION

β-chains of hemoglobin A is not known. My initial thought was that this is due to the loss of a slow point in the assembly of the β-chains occurring somewhere around position 92, as suggested by Winslow and Ingram (*J. biol. Chem.* **241**: 1144, 1966), and perhaps the loss of this position would speed up synthesis. However, in a recent paper by Clegg, et al. [*Nature (Lond.)*] **220**: 664, 1969], there was no evidence of a slow point in the assembly of either the α- or β-chains, so that I don't know what the explanation is.

E. RACHMILEWITZ (*Israel*): Did you study the met form of hemoglobin Philly? Is it stable or is it unstable?

R. F. RIEDER: We have not studied methemoglobin Philly.

R. SCHNEIDER (*USA*): I think it is of interest that when the same amino acid change present in hemoglobin Zürich occurs in another position in a molecule, as in hemoglobin P, which has the same histidine to arginine substitution in β 117, there are no inclusion bodies nor does the patient have any clinical symptomatology. This position is possibly, not entirely without clinical effects because it may be in the neighborhood of the α-β contact. Unfortunately, we have not been able to study this patient in detail because she is uncooperative. But there are target cells, a reticulocytosis and minimal alterations in the red cells that indicate that there may also be some instability in this hemoglobin.

N. KOSOWER: Dr. Rieder, you said that glutathione content of this cell was normal. Was it normal after the finding of inclusion bodies or was it normal without incubation?

R. F. RIEDER: This determination was done without redox dye incubation. In hemoglobin Zürich cells, glutathione levels were normal, as was glutathione stability. This was also true for hemoglobin Philly cells. The cells from patients with hemoglobin Gun Hill had slightly lower glutathione levels and normal stability.

N. KOSOWER: One would expect, as suggested by Jacob, a loss of glutathione with the precipitated hemoglobins. The cells containing these precipitates are presumably "recognized" by the removing organ, such as the spleen, and sequestered there. The question is whether this is due to the Heinz body per se bulging through the membrane, or due to a direct injury to the membrane. It seems that in some cases the Heinz body does stick in the membrane and may render the cell rigid. On the other hand, in the situation presented by us, the Heinz bodies do not bind to the membrane, and a direct injury to the membrane is observed. It may thus be that in some cases of genetically determined instability of hemoglobin, an injury caused by an oxidant cannot be "absorbed" by the hemoglobin, so that it precipitates. An additional challenge might then directly affect the membrane, and this membrane lesion is the one responsible for the "recognition" of the cell by the removing organ.

Discussion of paper presented by Dr. H. M. Ranney.

E. KLEIHAUER: I have two questions. One concerns hemoglobins with higher oxygen affinity, especially hemoglobins with α-chain abnormalities. Do you think that during pregnancy, oxygen transport across the placenta is impaired, and would this be harmful to the fetus? The second question concerns the higher oxygen affinity of fetal hemoglobin. As you know, the differences in oxygen affinity between cord blood and adult blood are no longer demonstrable in dialyzed hemoglobin solutions. In a recent paper by de Verdier and Garby (*Scand. J. clin. Lab. Invest.* **23**: 149, 1969), it was suggested that the lower binding of 2,3-diphosphoglycerate (DPG) to hemoglobin F might be an explanation for the higher oxygen affinity. As far as I know, a similar mechanism is thought to be involved in the high oxygen affinity of hemoglobin Hiroshima. Will you please comment on this.

H. M. RANNEY: The binding of organic polyphosphates by most of the abnormal hemoglobins has not been studied. Human fetal blood has been recognized to have a higher oxygen affinity than human adult blood, but in hemoglobin solutions of 0.1 M inorganic phosphate buffers, the oxygen equilibria of hemoglobin F did not appear to differ greatly from those of hemoglobin A. The recent studies of Tyuma and Shimizu (*Arch. Biochem.* **129**: 404, 1969) showed that, in the absence of phosphate, hemoglobin A exhibited a higher oxygen affinity than did hemoglobin F in the physiological pH range. The dramatic decrease in oxygen affinity which accompanies the addition of 2,3-DPG to hemoglobin A was markedly reduced in hemoglobin F. Thus in the presence of organic phosphate, hemoglobin F had a higher oxygen affinity than hemoglobin A.

Different results concerning the possible binding of 2,3-DPG to oxyhemoglobin, and concerning the binding of ratios of DPG:deoxyhemoglobin, have been obtained in different laboratories. There appears to be no doubt that deoxyhemoglobin binds DPG, but whether 1 or 2 mole of DPG are bound per tetramer of deoxyhemoglobin A is less certain. We have observed the expected decrease in oxygen affinity when DPG was added to hemoglobin $M_{Hyde\ Park}$ and hemoglobin $M_{Milwaukee-1}$ but we have not studied the actual binding of DPG to these variants. It has been suggested that the oxygen affinity of hemoglobin Hiroshima is related to altered DPG binding, but I have seen no published data on this.

E. KLEIHAUER: May I make a further comment on oxygen affinity of fetal hemoglobin. In cases of hereditary persistence of fetal hemoglobin, the oxygen affinity is the same as in adult blood. If this is true, I doubt that fetal hemoglobin has a different binding to 2,3-DPG or that only 2,3-DPG is responsible for the differences in oxygen affinity.

H. M. RANNEY: The oxygen affinity of the whole blood of patients with hereditary persistence of fetal hemoglobin has been said to be the same as that of normal blood. The red cell DPG levels of these patients are being studied by Dr. Charache and co-workers, but I do not know their current data.

Besides the presence of hemoglobin F in red cells of the newborn, other differences exist between adult and newborn red cells: these differences might affect the O_2 equilibria of the whole blood. There is suggestive evidence in the pedigrees of families with hemoglobins exhibiting high oxygen affinity of increased numbers of stillbirths. The family with hemoglobin Yakima, a β-chain abnormality, had a large number of abortions, but the mother and possible carrier of Yakima could not be tested.

E. KLEIHAUER: Just a short comment on methemoglobin spectra of unstable hemoglobins. A variety of unstable variants which have been investigated in our laboratory showed, with the exception of hemoglobin Wien, an abnormal spectrum in the range between 580 and 510 nm, but not between 600 and 650 nm, when compared with normal methemoglobin A. In addition, we can confirm the findings of Dr. Rachmilewitz about the spectral abnormality of methemoglobin H. Since hemoglobin Barts gives exactly the same spectrum as hemoglobin H, we believe it is typical for tetramers containing only one type of polypeptide chain.

Another comment concerns the heat denaturation test as a screening for unstable hemoglobins. We prefer to perform the test at high temperature (65 to 70 C) using buffered cyanmethemoglobin instead of oxyhemoglobin. By this method we get more reliable results since the formation of the very heat labile methemoglobin is avoided. Furthermore, results are available within 1 to 3 min.

H. M. RANNEY: Don't hemoglobins, other than those associated with the congenital Heinz body anemias, precipitate in this test? Substitutions at α^{47}, e.g. hemoglobin Beilinson and hemoglobin Hasharon, are unstable at higher temperatures although they are not usually associated with clinical evidence of accelerated blood destruction.

E. KLEIHAUER: We have investigated different normal and abnormal hemoglobin variants by this method. For example, hemoglobin S and hemoglobin F are twice as unstable as hemoglobin A; also hemoglobin M variants, Hörlein as well as Boston types, are rather heat labile. I agree that instability of a hemoglobin as judged from an *in vitro* procedure does not necessarily mean that this hemoglobin is also unstable *in vivo*, producing Heinz bodies or causing a hemolytic disease.

H. M. RANNEY: In populations where abnormal hemoglobins are frequently found, an electrophoretic screening method is generally used. The heat stability test, as you have described it, would seem to be an additional screening method. But the relationship of a positive test to a hemolytic state might not be as certain as in the original test of Dacie.

E. KLEIHAUER: You are right, heat denaturation is an additional screening method.

J. C. KAPLAN (*France*): I would like to ask Dr. Rachmilewitz three questions about his very interesting observations. First of all, have you any idea about the number of subunits which are present in the hemichromes? I mean are they monomers, dimers or polymers? Secondly, can you reverse the phenomenon at a certain stage by chemical means? And thirdly, do you think it could be possible to inject hemichromes *in vivo* and to study their relation to that black pigment which is present in the urine of patients with unstable hemoglobins?

E. RACHMILEWITZ: First I want to emphasize that hemichromes deserve more attention in the field of hemoglobin research. It is well known that one type of hemichrome can be formed by the addition of sodium benzoate or sodium salicylate to ferrihemoglobin A. When we did sedimentation velocity studies on these hemichromes, together with Dr. R. Brich we found that the S value of salicylate hemichrome was 2.4 to 2.2, instead of the value of 4.4 which is the normal S value for a tetramer of hemoglobin A. These results indicate that salicylate hemichrome is not a tetramer. It is not clear yet whether this hemichrome is composed of dimers or a combination of monomers, dimers and trimers. The S value was not low enough to state that the molecule was split into monomers only. Our working hypothesis is that sodium salicylate as well as sodium benzoate enter into the hemoglobin molecule so that the nonpolar aromatic rings disturb the tertiary structure of the subunit, thereby inducing changes in interchain interactions and dissociation of subunits, thus permitting hemichrome formation.

It was difficult to measure the S values of hemichromes since they tend to precipitate with time. We have investigated the possibility of reducing hemichromes, first by oxidizing hemoglobin H for 45 min and then by adding sodium ascorbate to the solution. The optical spectrum after 10 to 12 hr showed that almost all the hemoglobin was back in an oxy form. But we could not tell whether the reduced part of the solution came from ferrihemoglobin H or from hemichrome H. The problem was clarified using ESR spectroscopy. After 45 min we knew exactly how much of the solution remained as ferrihemoglobin and how much as hemichrome. After adding ascorbate and waiting for 12 hr, the repeated ESR recording showed that hemichrome was no longer present and only a small amount of ferrihemoglobin H was still left. ESR enables us to tell exactly the qualitative and quantitative differences between low spin (hemichromes) and high spin (ferrihemoglobin) compounds since the different absorption peaks appear in different regions of the magnetic field. We think that oxidation of hemoglobin H for longer than 45 min is followed by the formation of different types of hemichromes, measured by their g values, around $g=2$, which are not reducible.

The last proposition of Dr. Kaplan is most interesting. Again it is not so easy to get hemichromes in a stable form for a long time and to inject them into experimental animals.

N. KOSOWER: You showed very nicely that the precipitated inclusion bodies in hemoglobin H disease and in β-thalassemia consist of excess chains bound to heme in the form of hemichromes. Could you comment on such a finding in unstable hemoglobin? How would you reconcile it with the findings of Jacob, who claims that heme is not bound to the precipitated hemoglobin, for example, in hemoglobin Köln disease?

E. RACHMILEWITZ: The answer to your question, Dr. Kosower, is to try and look for the characteristics of hemichromes in other unstable and abnormal hemoglobins besides thalassemia. It could be that formation of hemichromes in hemoglobin Köln is an earlier step, preceding the separation of heme from globin. Dr. Lehmann promised to send us some of this hemoglobin and we will try to oxidize it and see what happens. At the present time, all that we can say is that in thalassemia the formation of hemichromes is the mechanism explaining the appearance of inclusion bodies. Hemichromes were also found in the oxidized form of hemoglobin Riverdale-Bronx $B6^{gly-arg}$. Whether this mechanism is common to all Heinz body hemolytic anemias is not known at present.

SESSION III

Chairmen: H. M. Ranney, *USA*
A. Szeinberg, *Israel*

Participants: R. Bochkowsky, *Israel*
N. Calmanovici, *Israel*
K. Fagelman, *USA*
L. Fogel, *USA*
E. Hegesh, *Israel*
J. C. Kaplan, *France*
M. Lupo, *Israel*
P. S. Paress, *USA*
J. M. Ross, *USA*
J. M. Schwartz, *USA*

DIAPHORASES OF HUMAN ERYTHROCYTES

EMANUEL HEGESH, NICU CALMANOVICI, MEIR LUPO and RUTH BOCHKOWSKY

Biochemical Research Laboratory, Kaplan Hospital, Rehovot, Israel

It is generally accepted that the hemoglobin in RBC is maintained in its reduced, active state by biological reducing mechanisms requiring enzymes which exhibit diaphorase activity. Several "diaphorases," otherwise called "methemoglobin reductases," have been isolated from human RBC. Some were found to be reduced nicotinamide adenine dinucleotide (NADH)-dependent (1–4), others to be reduced nicotinamide adenine dinucleotide phosphate (NADPH)-dependent (5–7). Although the diaphorases isolated to date promote the reduction of dyes such as 2,6-dichlorophenol-indophenol (DPIP) or methylene blue at high rates, almost all were quite ineffective in reducing methemoglobin directly on the addition of the corresponding reduced pyridine nucleotide. Several possibilities may explain these findings: a) A factor or a structural arrangement necessary for methemoglobin reduction, but not for diaphorase activity, is lost during purification. b) Reduction of methemoglobin is catalyzed by diaphorases other than those isolated. c) Methemoglobin prepared from nitrite-treated RBC is not identical with the physiological methemoglobin and is therefore not the proper substrate for the assay of "methemoglobin reductases."

Recently some electrophoretic variants of diaphorase, associated with cases of congenital enzymopenic methemoglobinemia, have been described (8,9). Furthermore, in analogy with the polymorphism of other RBC enzymes, the existence of unknown diaphorases or diaphorase isoenzymes in normal human RBC could not be excluded. This paper describes the electrophoretic characterization of a number of NADH- and NADPH-dependent diaphorases extracted from human RBC. Data are presented which suggest that diaphorase activity and methemoglobin reduction in human erythrocytes should be considered separately.

MATERIALS AND METHODS

Diethylaminoethyl (DEAE)-cellulose (Whatman-DE 11) was a product of W. & Balston Ltd., England. Acrylamide (Cyanogum 4l, gelling agent) and N,N,N[1],N[1]-tetramethylethylenediamine (TMED) were supplied by E.C. Apparatus Corp., USA. NADH (grade II), NADPH (type II) and DPIP sodium salt (grade I) were purchased from the Sigma Chemical Corp., USA. 3 (4,5-dimethyl thiazoly-2) 2,5 diphenyl 2H tetrazolium bromide (MTT) was a product of Nutritional Biochemical Corp., USA. Amido black 10 B and glycine were products of E. Merck, AG., Germany. All other chemicals used were of analytical grade.

Preparation of hemolysate. Three vol of blood were mixed with 1 vol of ACD solution (Citric acid, 0.8 g; sodium citrate, 2.2 g; dextrose, 2.45 g in 100 ml distilled water) and stored for not more than one week before use. The RBC were separated by centrifugation and washed three times with 0.9% Nacl. One vol of cells, packed by centrifugation, was hemolyzed by the addition of 1 vol of water and 0.4 vol of toluene. The mixture was shaken vigorously and centrifuged at high speed. The stroma and the toluene layer were aspirated and discarded. The hemolysate was then filtered through a Whatman No. 1 filter paper.

Preparation of the crude enzyme extract. DEAE cellulose was purified as described by Hennessey (10), except that the final washings were repeated until the pH of the ion exchanger was 6.8 to 7.0. An 8% suspension of the cellulose in water was prepared and added to the clear hemolysate in a ratio of 1/1 (v/v). The mixture was allowed to stand at 4 C for 1 hr, and stirred gently every 5 min. Under these conditions a great number of enzymes, comprising all diaphorases, are adsorbed on the cellulose, while hemoglobin remains in solution. The adsorbent was freed of residual hemoglobin by washing it three times with 1 mM phosphate buffer, pH 7.0, and then once with 0.3 mM phosphate buffer, pH 7.0. The diaphorases were eluted from the cellulose by a solution of 65 mM KCl, 5 mM citrate buffer, pH 5.2 and 50 μM EDTA. For each ml of hemolysate originally employed, 1 ml eluent was used. The suspension was stirred magnetically for 1 hr at 4 C and centrifuged. The supernatant fluid was collected by centrifugation at $2,000 \times g$ and concentrated 15 to 20 times by dialysis under reduced pressure at 4 C. It will be referred to as "concentrated crude enzyme extract." The extracts obtained from adult blood were yellow to brown in color and usually contained 300 to 400 mg protein/100 ml. Extracts from cord blood were more reddish in color and contained 1,000 to 1,500 mg protein/100 ml.

NADH-"ferrihemoglobin reductase" was assayed by the method of Hegesh et al. (4,11).

Disk electrophoresis was performed by the methods of Orenstein (12) and Davis (13) with the spacer gel omitted. The apparatus of the Shandon Scientific Company, London, was used. Two vol of the concentrated enzyme extract were mixed with 3 vol of 40% (w/v) sucrose. and 0.3 ml of the mixture was applied on top of the vertical column which contained 7.5% persulfate polymerized Cyanogum 41. Electrophoresis was run for 1 hr at 4 C at a constant current of 4 ma per column.

Specific staining for diaphorases. The method of Kaplan and Beutler (9) was used with a slight modification. It is based on the ability of diaphorases to promote the reduction of DPIP by reduced pyridine nucleotides. The reduced DPIP in turn transfers electrons nonenzymatically to MTT, causing the formation of an insoluble purple-blue formazan. Staining was performed for 30 min at 37 C in the dark. The staining solution contained per 3 ml: 2.7 μmole NADH or NADPH, 2.4 μmole MTT, 12.5 μmole Tris-HCl buffer, pH 7.55, 1.0 μmole EDTA and 0.06 μmole DPIP.

Staining for proteins. The gels were stained for 1 hr with 1% amido black in 7% acetic acid and decolorized electrophoretically using 2% acetic acid.

Purification of NADH- and NADPH-diaphorases. Purified enzyme preparation was achieved by gel filtration through a 3.1 × 100 cm column of packed G-100 Sephadex. The gel was equilibrated against a solution of 0.5 M sodium chloride in a 0.015 M sodium citrate buffer, pH 6.2. It was loaded with 3.5 ml of the crude concentrated enzyme extract containing 10 to 15 mg protein. Elution was performed with the same buffered sodium chloride solution at a flow rate of 50 ml/hr. Fractions of 3.5 ml were collected and stored in ice. The best fractions, constituting a 1,500-fold purification of the enzymes, were concentrated by dialysis against reduced pressure to a protein content of approximately 20 mg/100 ml and analyzed electrophoretically.

RESULTS

Fig. 1 demonstrates the electrophoretic patterns of the human RBC diaphorases. It can be seen that by using NADH as the electron donor, five colored bands are observed, while with NADPH seven or eight diaphorases appear. However, in most extracts, only the four upper bands, referred to as NADH 1 (D1); NADH 2 (D2); NADH 3 (D3) and NADH 4 (D4) could be demonstrated, while no variation was observed in the pattern of the NADPH-diaphorases. The third column was stained with amido black

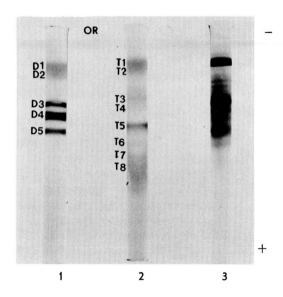

FIG. 1. NADH- and NADPH-diaphorases after polyacrylamide gel electrophoresis. The columns were loaded with concentrated crude enzyme extract from normal human RBC and stained specifically for diaphorase (see Materials and Methods). In this photograph, the migration of the enzymes is downward, the anode being at the bottom. Column 1 was stained with NADH, column 2 with NADPH and column 3 with amido black. Or = Origin.

for proteins and is shown for comparison. The two upper bands in columns 1 and 2 seem to be due to NADH-NADPH cross activity.

In order to define the contribution of each of the components of the staining mixture to the electrophoretic pattern, we prepared a number of identical gel columns and stained each one differently, by omitting one of the ingredients of the staining mixture in turn. Results are shown in Fig. 2. Columns 1 to 5 were stained for NADH-diaphorases. Column 1 demonstrates the pattern obtained with the complete system. Column 2 was stained with a mixture in which pyridine nucleotides were omitted, while column 3 was stained in the absence of diazonium salt. It is evident that in the absence of either reduced pyridine nucleotide or diazonium salt the staining reaction did not proceed. The pale bands on the top of columns 2, 3 and 4 were orange-red in color and probably due to a heme compound contaminating the enzyme. In column 5, where the enzyme was omitted, this contaminating band was absent. Column 4 was stained in the absence of DPIP. It can be seen that some of the diaphorases, especially NADH 3 (D3), transferred

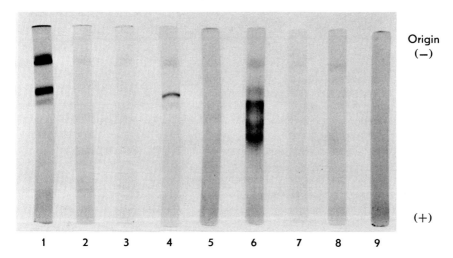

FIG. 2. Component study of the incubation media for the staining of NADH- and NADPH-diaphorases after polyacrylamide gel electrophoresis.
In columns 1,3 to 5 reduction was initiated by NADH; in columns 6 to 9, by NADPH. The components omitted were: in 1 and 6 none, in 2 both pyridine nucleotides, in 3 and 7 MTT, in 5 and 9 the enzyme.

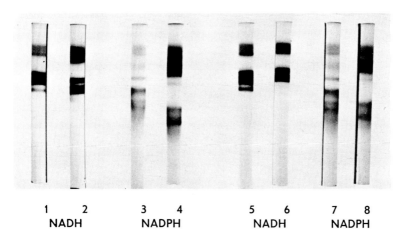

FIG. 3. Electrophoretic patterns of the diaphorases of normal and cord blood.
Columns 1, 3, 5 and 7 represent the pattern of enzyme extracts from normal adult blood; columns 2, 4, 6 and 8 extracts from cord blood. Columns 1 to 4 were loaded to contain the same total NADH-ferrihemoglobin reductase activity. In columns 5 to 8 the same volume of extract, obtained under identical conditions, was used.

electrons directly from NADH to MTT. However, in the presence of DPIP this transfer proceeded at a much higher rate. When incubation time was extended, all diaphorases could be made visible in the absence of DPIP. Columns 6 to 9 were stained for NADPH-diaphorases. Column 6 represents the normal pattern. Omitting any of the components of the staining mixture resulted in the disappearance of the colored bands. Thus, although diaphorases are known to promote the reduction of diazonium salts directly by reduced pyridine nucleotides (14), it seems that DPIP is necessary in this system for the efficient reduction of MTT.

It has been shown by several investigators that in comparison to normal adult blood, blood from umbilical cords or from newborn infants exhibits a low NADH-diaphorase activity (15–17). This deficiency is supposed to play a role in the transitory methemoglobinemia observed in newborns exposed to oxidizing agents. It was therefore considered of interest to characterize the diaphorases of cord blood electrophoretically. Fig. 3

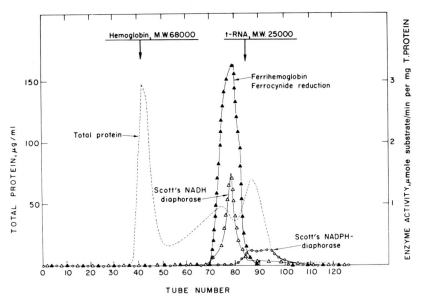

FIG. 4. Elution profile of NADH and NADPH diaphorases separated on a Sephadex G-100 column.
For technique see Materials and Methods. Activity of NADH- and NADPH-diaphorase was determined according to Scott et al. (1) and of ferrihemoglobin-ferrocyanide reductase by the method of Hegesh et al. (4). Total protein was measured by the method of Lowry et al. (18). Collection of fractions was started after placing the crude enzyme extract on the top of the column.

shows a comparision of the electrophoretic patterns of the diaphorases of adult and cord blood.

In columns 1 to 4, amounts of extract with the same total methemoglobin reductase activity were applied. Columns 5 to 8 were loaded with an equal volume of extract, obtained under identical conditions, allowing a comparison of the approximate concentrations of the different diaphorases in both types of RBC. The NADH-diaphorases of adult blood (columns 1, 3, 5, 7) as compared to cord blood (columns 2, 4, 6, 8) seem to be similar in electrophoretic mobility. However, a band in adult blood NADPH 5 (T5) was regularly absent in cord blood. Furthermore, there is evidently more NADPH-diaphorase activity in cord blood than in adult blood (compare column 3 with 4 and column 7 with 8). It seems, therefore, that the enzyme pattern of adult blood is slightly different from that of cord blood.

We extended our investigation by applying polyacrylamide gel electrophoresis to the study of two diaphorases purified from human RBC. A 1,500-fold purification was achieved by gel filtration on a Sephadex G-100 column. Fig. 4 demonstrates the elution profile of a typical experiment. It can be seen that the main protein, which could be shown to consist mainly of hemoglobin, was eluted first. NADH-diaphorase, measured by Scott's

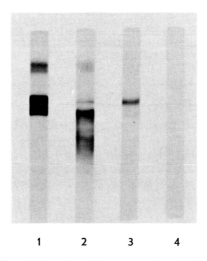

FIG. 5. Polyacrylamide gel electrophoresis patterns of purified NADH-diaphorase. Columns 1 and 2 represent the patterns of the NADH- and NADPH-diaphorases from normals used in this experiment as controls. Columns 3 and 4 were loaded with purified NADH-diaphorase (fraction I). They were stained with NADH (column 3), as well as with NADPH (column 4).

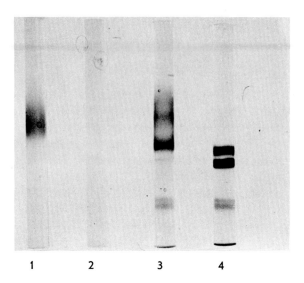

FIG. 6. Polyacrylamide gel electrophoresis pattern of purified NADPH-diaphorase. Columns 1 and 2 represent the patterns of the NADH- and NADPH-diaphorases from normal blood, used as controls. Columns 3 and 4 were loaded with purified NADPH-diaphorase (fraction II) and stained with NADH (column 3), as well as with NADPH (column 4).

dichlorophenol-indophenol method as modified by Ross (17), or by our methemoglobin-ferrocyanide method was eluted between the second and third protein peaks. A NADPH-diaphorase, almost free of NADH-diaphorase activity, was eluted next (Fig. 4). Considering the molecular weights of hemoglobin and transfer-RNA, which were used as markers, the molecular weight of the NADH-diaphorase could be estimated as approximately 30,000, while that of the NADPH-diaphorase, as 20,000 (Fig. 4). The electrophoretic patterns of the two diaphorases are shown in Fig. 5,6. The NADPH-diaphorase (Fig. 5) resolved into two bands, corresponding to diaphorases, NADH 3 (D3) and NADH 4 (D4) of the crude extract. No NADPH-diaphorase activity could be detected in this fraction. The NADPH diaphorase exhibited three of four bands and by this method had an absolute specifity for NADPH (Fig. 6).

Table 1 summarizes the specificity of the purified diaphorases for different substrates. As can be seen, the NADH-diaphorase (fraction I) is specific for NADH in Scott's and in our ferrihemoglobin-ferrocyanide assay system (Table 1, assay systems 1 and 2). However, when using cytochrome C as the final electron acceptor (system 3) the enzyme was also

TABLE 1. *Specificity of the purified RBC diaphorases*

Assay system	Final electron acceptor	Activator	Pyridine nucleotide	Fraction I (Tubes 75 to 80)	Fraction II (Tubes 85 to 98)
			μmole	Substrate/min	mg protein X 10^3
1	2,6-dichlorophenol-indophenol	None	NADH	1,500	6
			NADPH	0	65
2	Ferrihemoglobin	Ferrocyanide	NADH	3,400	95
			NADPH	0	0
3	Cytochrome C	None	NADH	15,400	450
			NADPH	1,500	900
4	Ferrihemoglobin	None	NADH	0	0
			NADPH	0	0
5	Ferrihemoglobin	Methylene blue	NADH	833	60
			NADPH	1,190	103

Assay 1 was performed by the diaphorase method of Scott et al. (1).
In assays 2, 3 and 4, the method of Hegesh and Avron (4) was used with the following modifications: In assay 3, ferrihemoglobin was replaced by ferricytochrome C at a concentration of 2 μmole/ml; in assays 3, 4 and 5, ferrocyanide was excluded from the reaction mixtures; in assay 4, 0.5 ml of 20 times concetrated crude enzyme extract was incorporated; in assay 5, ferrihemoglobin was used at a concentration of 0.1 μmole/ml and methylene blue was added at a final concentration of 0.3 μmole/ml.

active with NADPH (10% of its activity with NADH). Under the experimental conditions employed, neither enzyme fraction catalyzed the reduction of methemoglobin directly, in the absence of an activator or electron carrier (Table 1, system 4). In the presence of methylene blue (system 5), both fractions, the NADH-diaphorase and the NADPH-diaphorase, had a greater affinity for NADPH as compared to NADH. Therefore the possibility is suggested that the enzymes isolated by Kiese et al. (5), Huenneckens et al. (6) and Shimuzu et al. (7), which were measured in the presence of methylene blue and defined as NADPH-methemoglobin reductases, may have been identical with the NADH diaphorase isolated by Scott et al. (1) or with the ferrihemoglobin reductase isolated in our laboratory (4).

Fig. 7 demonstates the polyacrylamide gel electrophoresis pattern obtained in a case of "congenital enzymopenic methemoglobinemia." The total NADH-ferrihemoglobin reductase activity in this patient was found to be less than 10% of normal. It can be clearly seen that the RBC are normal in respect to NADPH-diaphorase (Fig. 7, column 4), while they are devoid

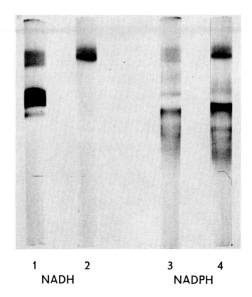

1 2   3   4
NADH   NADPH

FIG. 7. Polyacrylamide gel electrophoresis patterns of crude enzyme extract from RBC of a patient with congenital enzymopenic methemoglobinemia.
Columns 1 and 3 were loaded with the crude enzyme extract from normal adult blood and columns 2 and 4 with the extract from the patient's blood. The staining mixtures contained NADH or NADPH, as indicated in the Fig.

completely of the NADH 3 (D3) and NADH 4 (D4) and NADH 5 (D5) enzymes (Fig. 7, column 2). The NADH 1 (D1) and NADH 2 (D2) enzymes on top of the column seem to be identical with the residual "diaphorase" found by Scott in homozygotic patients with congenital methemoglobinemia due to diaphorase deficiency (19). The disorder in this patient can therefore be defined as a lack of NADH-diaphorases 3, 4 and 5.

DISCUSSION

It can be stated that the presence of a considerable number of NADH- and NADPH-dependent diaphorases, not described to date, was demonstrated in extracts from human RBC. The physiological significance of the different enzymes or isoenzymes, especially their function as methemoglobin reductases, remains to be determined. Differences were observed in the diaphorase pattern of cord blood as compared to adult blood. However, under the experimental conditions used, the deficiency of cord blood in total NADH-diaphorase activity could not be demonstrated electrophoretically. In a subject with congenital recessive methemoglobinemia, the NADH 3 (D3), NADH 4 (D4) and NADH 5 (D5) enzyme bands, which seem to be primarily responsible for the reduction of methemoglobin in RBC, were absent. It could be shown that the NADH-diaphorase preparation isolated in our laboratory seems to be identical with the NADH 3 (D3) and NADH 4 (D4) diaphorase bands. However, this preparation, although highly active as a NADH-diaphorase, was found inactive in catalyzing the direct reduction of methemoglobin upon addition of NADH. It is therefore, proposed that NADH-diaphorase activity and NADH-methemoglobin reductase activity should be considered separately. In additon to NADH and diaphorase, an unknown factor or a structural arrangement seems to be necessary for the functioning of the NADH-methemoglobin reduction system in human RBC.

The electrophoretic technique described here may be of help in the investigation of the genetic polymorphism of the diaphorases of human RBC and its connection with disorders of the methemoglobin reduction mechanism.

We are indebted to Dr. Ernst Jaffé, Department of Medicine, Albert Einstein College of Medicine, New York, for his helpful suggestions and for supplying us with a blood sample from a congenital methemoglobinemia patient.

REFERENCES

1. SCOTT, E. M., DUNCAN, J. W. and EKSTRAND, V. The reduced pyridine nucleotide dehydrogenases of human erythrocytes. *J. biol. Chem.* **240**: 481, 1965.
2. SCOTT, E. M. and MCGRAW, J. C. Purification and properties of diphosphopyridine nucleotide diaphorases of human erythrocytes. *J. biol. Chem.* **237**: 249, 1962.
3. KAJITA, A., KERWAR, G. K. and HUENNEKENS, F. M. Multiple forms of methemoglobin reductase. *Arch. biochem.* **130**: 662, 1969.
4. HEGESH, E. and AVRON, M. The enzymatic reduction of ferrihemoglobin. II. Purification of a ferrihemoglobin reductase from human erythrocytes. *Biochim. biophys. Acta (Amst.)* **146**: 397, 1967.
5. KIESE, M., SCHNEIDER, C. and WALLER, H. D. Hemiglobinreduktase, *Naunyn-Schmiedeberg's Arch. exp. Path. Pharmak.* **231**: 158, 1957.
6. HUENNEKENS, F. M., CAFFREY, R. W., BASFORD, R. E. and GABRIO, B. W. Erythrocyte metabolism. IV. Isolation and properties of methemoglobin reductase. *J. biol. Chem.* **227**: 261, 1957.
7. SHIMIZU, C. and MATSUURA, F. Purification and some properties of "methemoglobin reductase." *Agr. Biol. Chem.* **32**: 587, 1968.
8. WEST, C. A., GOMPERTS, B. D., HUEHNS, E. R., KESSEL, I. and ASHBY, J. R. Demonstration of an enzyme variant in a case of congenital methaemoglobinaemia. *Brit. med. J.* **2**: 212, 1967.
9. KAPLAN, J. C. and BEUTLER, E. Electrophoresis of red cell NADH and NADPH-diaphorases in normal subjects and patients with congenital methemoglobinemia. *Biochem. biophys. Res. Commun.* **29**: 605, 1967.

10. HENNESSEY, M. A., WALTERSDORPH, A. M., HUENNEKENS, F. M. and GABRIO, B.W. Erythrocyte metabolism. VI. Separation of erythrocyte enzymes from hemoglobin. *J. clin. Invest.* **41**: 1257, 1962.
11. HEGESH, E., CALMANOVICI, N. and AVRON, M. New method for determining ferrihemoglobin reductase (NADH-methemoglobin reductase) in erythrocytes. *J. Lab. clin. Med.* **72**: 339, 1968.
12. ORNSTEIN, L. Disk electrophoresis. I. Background and theory. *Ann. N.Y. Acad. Sci.* **121**: 321, 1964.
13. DAVIS, B.S. Disk electrophoresis. II. Method and application to human serum proteins. *Ann. N.Y. Acad. Sci.* **121**: 404, 1964.
14. BURSTONE, M. S. "Enzyme Hystochemistry." New York, Academic Press, 1962, p. 491.
15. MCDONALD, C. D. and HUISMAN, T. H. A comparative study of enzymic activities in normal adult and cord blood erythrocytes as related to the reduction of methemoglobin. *Clin. chim. Acta* **7**: 555, 1962.
16. BETKE, K., KLEIHAUER, E., GAERTHER, Ch. and SCHIEBE, G. Verminderung von Methamoglobinreduktion, Diaphoraseaktivität und Flavinen in Erythrozyten junger Säuglinge. *Arch. Kinderheilk.* **170**: 66, 1964.
17. ROSS, J. D. Deficient activity of DPNH-dependent methemoglobin diaphorase in cord blood erythrocytes. *Blood* **21**: 51, 1963.
18. LOWRY, O. H., ROSEBROUGH, N. J., FARR, A. L. and RANDALL, R. J. Protein measurement with the folin phenol reagent. *J. biol. Chem.* **193**: 265, 1951.
19. SCOTT, E. M. Purification of diphosphopyridine nucleotide diaphorase from methemoglobinemic erythrocytes. *Biochem. biophys. Res. Commun.* **9**: 59, 1962.

STUDIES ON NADH AND NADPH DIAPHORASES IN BLOOD CELLS FROM NORMAL SUBJECTS AND IN CONGENITAL METHEMOGLOBINEMIA

JEAN-CLAUDE KAPLAN

Institute of Molecular Pathology, Paris and School of Medicine, Rouen, France.

It is well established that, in the red cell, there are two diaphorase systems differing with respect to their specificity towards NADH and NADPH (1–12). These enzymes are sometimes called "methemoglobin reductases" because of their role in methemoglobin reduction in intact cells. A deficiency of the NADH-dependent system gives rise to congenital methemoglobinemia (1, 8, 13–15), whereas deficiency involving the NADPH-dependent system does not produce methemoglobinemia (16). The latter condition illustrates the lack of physiological significance of the NADPH diaphorase, an enzyme which is unable to reduce methemoglobin in intact red cells in the absence of an artificial dye such as methylene blue (3, 5, 7, 17–20). Chemical separation of the two systems has been performed by several groups of investigators (2, 6, 9–11). A system acting as a NADH-ferrocyanide methemoglobin reductase was also isolated (21, 22).

We have described a method for staining both diaphorases after starch gel electrophoresis. A clear separation of NADH and NADPH diaphorases was achieved by this method (23).

We will discuss here the results obtained after electrophoresis of both diaphorases in normal red cells, normal leukocytes and NADH-diaphorase deficient red cells. We will also present a simple method devised for fast detection of red cell NADH-diaphorase deficiency (24).

ELECTROPHORETIC DEMONSTRATION OF NADH AND NADPH DIAPHORASES IN BLOOD CELLS

The specific method of staining which we devised is based upon the fact that the reduced form of 2,6 dichlorophenolindophenol (DCIP), a

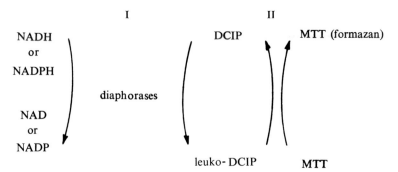

FIG. 1. DCIP-MTT linked method for revelation of NADH and NADPH diaphorase activity (23).

product of the reaction with both enzymes, can chemically reduce some tetrazolium salts into their insoluble formazan derivative (23) (Fig. 1).

During the process, oxidized DCIP is regenerated, therefore catalytical amounts of this dye are to be used in the staining system. The result is a positive purple band of formazan, clearer and more stable than such negative stains as those given by the decolorization of DCIP (25, 26) or the defluorescence of the reduced pyridine nucleotide. In the absence of DCIP, the diaphorase systems are unable to promote direct electron transfer from NADH or NADPH to tetrazolium derivatives.

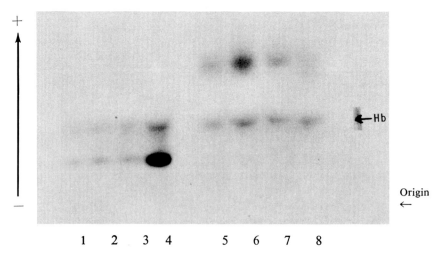

FIG. 2. Electrophoresis of red cell diaphorases (23): 1–4) NADH diaphorase (slowest bands), 5–8) NADPH diaphorase (fastest bands).

The electrophoretic studies were carried out in starch gel at pH 9.6 with a voltage gradient of 20 to 25 v/cm (23). After 4 hr, the sliced gel was overlayed with a mixture of NADH (or NADPH) 1.3 mM, DCIP 0.06 mM, MTT 1.2 mM in a Tris-HCl 0.25 M buffer, pH 8.4. The diaphorase bands became visible as blue purple bands, after incubation in the dark at room temperature.

Red and white cells from normal humans, and red cells from subjects with congenital methemoglobinemia were studied. The electrophoretic pattern observed depends on the nature of the reduced pyridine nucleotide which is used in the staining system.

Normal red cell diaphorases. With NADH, one single band appeared behind hemoglobin A, at the same level as hemoglobin A_2, representing the NADH diaphorase (Fig. 2, 3). Hemolysates freed from hemoglobin by treatment with DEAE cellulose (27) and concentrated by vacuum dialysis gave the same pattern.

In "aged" samples, stored several days at + 4 C or frozen for several weeks, multiple bands were seen; in crude hemolysates, a second more anodic band appeared just behind hemoglobin A (Fig. 3). In "aged" hemoglobin-free concentrate sample, the original single band split into three bands (see Fig. 7). Treatment by β-mercaptoethanol did not reverse the pattern. This multiple banding phenomenon might represent a denaturation of the enzyme since it was not observed in fresh preparations.

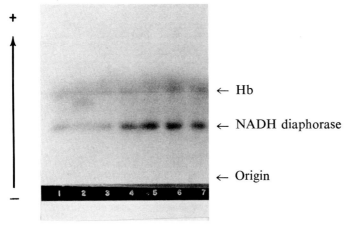

FIG. 3. Red cell NADH-diaphorase electrophoresis: 1, 3–7) fresh hemolysates, 2) "aged" hemolysate (stored frozen three weeks at –20 C). Notice second NADH-diaphorase band just behind the Hb band.

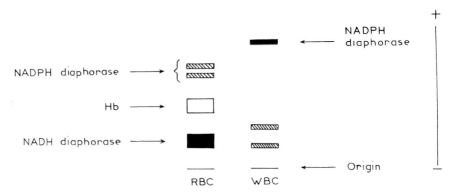

FIG. 4 Diagram of the electrophoretic pattern of NADH and NADPH diaphorases in normal RBC and WBC.

With NADPH, a broader zone was stained far ahead of hemoglobin A. It corresponded to the NADPH-diaphorase activity appearing as a double band, not clearly resolved (Fig. 2, 4).

Among 200 subjects studied for red cell NADH diaphorase, no electrophoretic polymorphism was detected (23), a finding which of course does not preclude the existence of rare variants. In one subject, the red cell NADPH diaphorase was found to have an abnormally fast mobility (J. C. Kaplan and E. Beutler, unpublished data). This case most probably represents an electrophoretic variant of the NADPH diaphorase. Unfortunately, it could not be investigated any further.

The red cell NADH-diaphorase electrophoretic pattern was found to be normal in 20 cord blood samples (23). Therefore, the decreased NADH-diaphorase activity which is observed in newborns (28) did not appear to be related to a fetal form which can be demonstrated by electrophoresis.

The red cell NADPH-diaphorase electrophoretic phenotype is identical in Caucasians and Negroes, regardless of their glucose-6-phosphate dehydrogenase phenotype (23). This rules out the possibility of a structural relationship between the two enzymes, a hypothesis which had been put forward to explain the decrease of NADPH diaphorase often found in glucose-6-phosphate dehydrogenase-deficient red cells (19, 29).

Normal leukocytes. The electrophoretic pattern of diaphorases is somewhat different in these cells (A. Hanzlickova-Leroux and J. C. Kaplan, unpublished data). Preliminary experiments have shown that a NADPH diaphorase is readily released in the supernatant fraction of lysed leukocytes. Upon electrophoresis it appeared as a single sharp and strong band, faster than the red cell NADPH diaphorase (Fig. 4). In contrast, the NADH diaphorase did not appear on the gel unless a total disruption of subcellular structures had been achieved by repeated freezing and thawing or sonication. At least two bands of equal intensity were stained after electrophoresis, one band having the same mobility as the single NADH-diaphorase band found in red cells (Fig. 4). The significance of these data is currently being investigated.

Red cell diaphorase electrophoresis in congenital recessive methemoglobinemia. We have studied the electrophoretic pattern of red cell diaphorases in several cases of NADH-diaphorase deficiency. Including our first reported case (23), a total of six homozygotes were investigated.

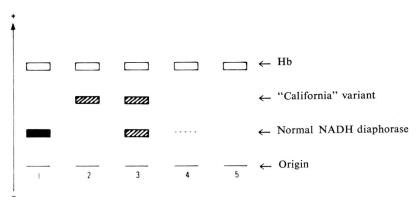

FIG. 5. Diagram of the electrophoretic pattern observed in the red cells of NADH-diaphorase deficient subjects. 1) Normal, 2) "California" homozygote, 3) "California" heterozygote, 4) deficient homozygote with weak enzyme activity and normal enzyme mobility ("type II" in text), 5) deficient homozygote without any detectable activity ("type I" in text).

TABLE 1. *Personal results of NADH-diaphorase electrophoresis in congenital methemoglobinemia*

No band visible	Faint band with normal mobility	Weak band with modified mobility	
3 subjects * (homozygotes)	2 subjects (homozygotes)	1 homozygote 1 heterozygote	Type "California"

* One with mental retardation

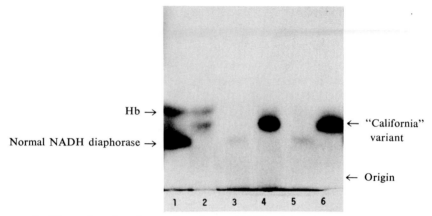

FIG. 6. Electrophoresis of normal and type "California" NADH diaphorases (23): 1) Normal crude hemolysate, 2) type "California" deficient hemolysate, 3, 5) normal hemoglobin-free hemolysate (DEAE treated), 4, 6) type "California" deficient hemoglobin-free hemolysate (DEAE treated).

The results are summarized as follows (Fig. 5) (Table 1):

1) In three cases, the NADH-diaphorase band was completely missing in both crude hemolysates and hemoglobin-free concentrated preparations.

2) In two cases, a very weak NADH-diaphorase band retaining a normal mobility was seen in crude hemolysates (Fig. 5, channel 4). It was no longer visible in hemoglobin-free concentrated samples.

In both groups the NADH-diaphorase residual activity in the red cells was very low (less than 10%) as usually described. In the heterozygous relatives the electrophoretic pattern of the enzyme was normal.

3) In one family a homozygote was found to have an abnormal, fast moving NADH diaphorase located between hemoglobin A and A_2 (Fig. 5, 6).

The band was weak in crude hemolysates but stained strongly in hemoglobin-free concentrated samples. The normal band was not detectable. This subject, an adult Caucasian female with a typical case history of congenital methemoglobinemia, had an unusually high residual NADH-diaphorase activity in her red cells, approximately 30% of the normal level. This fast moving enzyme variant was named the "California variant" (23). It is noteworthy that "aged" extracts of this variant displayed, after electrophoresis, a three banded pattern differing from the normal "aged" pattern, as indicated on the diagram (Fig. 7).

In the hemolysate, as well as in hemoglobin-free concentrated samples of the subject's daughter red cells, two NADH-diaphorase bands were seen:

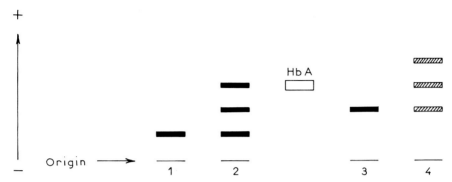

FIG. 7. Diagram of NADH-diaphorase electrophoresis of fresh and "denatured" hemoglobin-free extracts. 1) Fresh extract of normal RBC, 2) "aged" extract of normal RBC, 3) fresh extract of "California" variant RBC, 4) "aged" extract of "California" variant.

one with normal mobility, the other located at the same place as the "California" variant. The presence of both alleles in the heterozygote is a further indication of the validity of the method.

It should be emphasized that in all these cases the NADPH-diaphorase electrophoretic pattern remained unchanged. Conversely in his unique case of NADPH-diaphorase deficiency, Sass found the NADPH-diaphorase band to be missing (M. D. Sass, personal communication).

By using the DCIP decolorization technique, West and associates have also described an electrophoretically detectable NADH diaphorase with abnormal fast mobility in one subject with congenital methemoglobinemia (26). This variant however has a different mobility from the "California" variant (E. R. Huehns, personal communication).

Congenital recessive methemoglobinemia now appears to be heterogeneous from a molecular point of view. At least three different types could be tentatively defined on the basis of electrophoresis (Table 1).

Type I refers to the three cases without any NADH-diaphorase activity detectable upon electrophoresis. Type II refers to the two cases in which a weak activity with normal mobility was detectable in crude hemolysates. The activity was lost in the hemoglobin-free concentrated samples, suggesting lability of the enzyme. Type III refers to the cases in which the NADH diaphorase displays an abnormal electrophoretic mobility. It includes the "California" variant (23), the variant described in London (26), and any other still undescribed electrophoretic variant with low but detectable activity.

A SIMPLE METHOD FOR FAST DIAGNOSIS OF NADH-DIAPHORASE DEFICIENCY (UV SPOT-TEST METHOD)

The diagnosis of NADH-diaphorase deficiency can be achieved by any of the following methods: measurement of the rate of methemoglobin reduction in intact red cells incubated with glucose, lactate or inosine (4, 5, 19); estimation of the enzyme activity in hemolysates using the NADH-DCIP reductase method of Scott (15) or the NADH-ferrocyanide methemoglobin reductase method of Hegesh and associates (30, 31).

We now present a semiquantitative method suitable for fast screening and diagnosis of NADH-diaphorase deficiency (24). It was devised in order to provide an answer within 1 hr, without need for any special reagents or expensive equipment. Our method is based on the UV spot-test designed by Beutler for the fast detection of various red cell enzyme activities (32, 33).

A small amount of whole blood is incubated in a lysing medium containing NADH and DCIP (Fig. 8). Spots are made on filter paper at 5 min intervals and allowed to dry. Upon examination under long wave UV light, a complete defluorescence can be observed with normal blood in less than 20 min. It is due to NADH consumption and is an index of the NADH-diaphorase activity.

Four homozygous subjects with NADH-diaphorase deficiency were investigated by this method. All displayed a significant increase in the deflu-

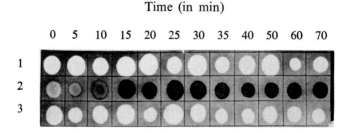

FIG. 8. UV spot-test for NADH diaphorase. Experimental conditions (24): Preincubation of whole blood with sodium nitrite (final concentration, 9 mM), during 30 min. Incubation of 20 µl of unwashed nitrited blood at 37 C in a mixture made of 240 µl of a 0.06 M Tris-HCl buffer, pH 7.6, containing 0.4 µg saponin, 0.14 µmole NADH, 0.038 µmole DCIP and 0.054 µmole EDTA. 1) Reaction mixture alone, 2) reaction mixture + normal blood, 3) reaction mixture + NADH-diaphorase deficient blood (homozygous subject).

orescence time (Fig. 8). Therefore, we consider that the UV spot-test is of good value when a fast diagnosis is needed, or for screening.

However, since the results obtained with heterozygous subjects were not clearcut, we do not recommend it for the detection of heterozygotes.

REFERENCES

1. GIBSON, Q. H. The reduction of methaemoglobin in red cells and studies on the cause of idiopathic methaemoglobinaemia. *Biochem. J.* **42**: 13, 1948.
2. HUENNEKENS, F. M., CAFFREY, R. W., BASFORD, R. E. and GABRIO, B. W. Erythrocyte metabolism. IV. Isolation and properties of methemoglobin reductase. *J. biol. Chem.* **227**: 261, 1957.
3. JAFFE, E. R. Metabolic processes involved in the formation and reduction of methemoglobin in human erythrocytes, in: Bishop, C. and Surgenor, D. M. (Eds.), "Red blood cells." New York, Academic Press, 1964, p. 397.
4. JAFFE, E. R. DPNH-methemoglobin reductase (diaphorase), in: Yunis, J. J. (Ed.), "Biochemical methods in red cells genetics." New York, Academic Press, 1969, p. 231.
5. KAPLAN, J. C. Les systemes d'oxydo-réduction du globule rouge et leurs anomalies. *Nouv. Rev. franc. Hémat.* **6**: 809, 1966.
6. KIESE, M., SCHNEIDER, C. and WALLER, H. D. Hämiglobin reduktase. *Naunyn-Schmiedeberg's Arch. exp. Path. Pharmak.* **231**: 158, 1957.
7. ROSS, J. D. and DESFORGES, J. F. Erythrocyte glucose-6-phosphate-dehydrogenase activity and methemoglobin reduction. *J. Lab. clin. Med.* **54**: 450, 1959.
8. SCOTT, E. M. and Griffith, I. V. The enzymic defect of hereditary methemoglobinemia: diaphorase. *Biochim. biophys. Acta (Amst.)* **34**: 584, 1959.
9. SCOTT, E. M. and MC GRAW, J. C. Purification and properties of diphospho-pyridine nucleotide diaphorase of human erythrocytes. *J. biol. Chem.* **237**: 249, 1962.
10. SCOTT, E. M., DUNCAN, I. W. and EKSTRAND, V. The reduced pyridine nucleotide dehydrogenases of human erythrocytes. *J. biol. Chem.* **240**: 481, 1965.
11. SHRAGO, E. and FALCONE, A. B. Human erythrocyte reduced triphospho-pyridine nucleotide oxidase. *Biochim. biophys. Acta (Amst.)* **67**: 147, 1963.
12. SMITH, R. P. The oxygen and sulfide binding characteristics of hemoglobin generated from methemoglobin by two erythrocytic systems. *Molec. Pharmacol.* **3**: 378, 1967.
13. JAFFE, E. R. and HELLER, P. Methemoglobinemia in man, in: Moore, C. V. and Brown, E. B. (Eds.), "Progress in hematology." New York, Grune and Stratton, Inc., 1964, v. 4, p. 48.
14. JAFFE, E. R. Hereditary methemoglobinemias associated with abnormalities in the metabolism of erythrocytes. *Amer. J. Med.* **41**: 786, 1966.
15. SCOTT, E. M. The relation of diaphorase of human erythrocytes to inheritance of methemoglobinemia. *J. clin. Invest.* **39**: 1176, 1960.
16. SASS, M. D., CARUSO, C. J. and FARHANGI, M. TPNH methemoglobin reductase deficiency: a new red cell enzyme defect. *J. Lab. clin. Med.* **70**: 760, 1967.
17. BEUTLER, E. and BALUDA, M. C. Methemoglobin reduction. Studies of the interaction between cell populations and of the role of methylene blue. *Blood* **22**: 323, 1963.
18. JAFFE, E. R. The reduction of methemoglobin in erythrocytes incubated with purine nucleosides. *J. clin. Invest.* **38**: 1555, 1959.
19. JAFFE, E. R. The reduction of methemoglobin in erythrocytes of a patient with congenital methemoglobinemia, subjects with erythrocyte glucose-6-phosphate-dehydrogenase deficiency and normal individuals. *Blood* **21**: 561, 1963.
20. SASS, M. D., CARUSO, C. J. and AXELROD, B. R. Mechanism of the TPNH-linked reduction of methemoglobin by methylene blue. *Clin. chim. Acta* **24**: 77, 1969.

21. HEGESH, E. and AVRON, M. The enzymatic reduction of ferrihemoglobin. I. The reduction of ferrihemoglobin in red blood cells and hemolysates. *Biochim. biophys. Acta (Amst.)* **146**: 91, 1967.
22. HEGESH, E. and AVRON, M. The enzymatic reduction of ferrihemoglobin. II. Purification of a ferrihemoglobin reductase from human erythrocytes. *Biochim. biophys. Acta (Amst.)* **146**: 397, 1967.
23. KAPLAN, J. C. and BEUTLER, E. Electrophoresis of red cell NADH- and NADPH-diaphorases in normal subjects and patients with congenital methemoglobinemia, *Biochem. biophys. Res. Commun.* **29**: 605, 1967.
24. KAPLAN, J. C., NICOLAS, A. M., HANZLICKOVA-LEROUX, A. and BEUTLER, E. A simple spot screening test for fast detection of red cell NADH-diaphorase deficiency. *Blood* (in press).
25. BREWER, G. J., EATON, J. W., KNUTSEN, C. S. and BECK, C. C. A starch gel electrophoretic method for the study of diaphorase isozymes and preliminary results with sheep and human erythrocytes. *Biochem. biophys. Res. Commun.* **29**: 198, 1967.
26. WEST, C. A., GOMPAERTS, B. D., HUEHNS, E. R., KESSELL, I. and ASHBY, J. R. Demonstration of an enzyme variant in a case of congenital methaemoglobinaemia. *Brit. med. J.* **4**: 212, 1967.
27. HENNESSEY, M. A., WALTERSDORPH, A. M., HUENNEKENS, F. M. and GABRIO, B. W. Erythrocyte metabolism. VI. Separation of erythrocyte enzymes from hemoglobin. *J. clin. Invest.* **41**: 1257, 1962.
28. ROSS, J. D. Deficient activity of DPNH-dependent methemoglobin diaphorase in cord blood erythrocytes. *Blood* **21**: 51, 1963.
29. BONSIGNORE, A., FORNAINI, G., SEGNI, G. and FANTONI, A. Glutathione reductase in erythrocytes of human subjects with a case history of favism. *Ital. J. Biochem.* **9**: 345, 1960.
30. HEGESH, E., CALMANOVICI, N. and AVRON, M. New method for determining ferrihemoglobin reductase (NADH-methemoglobin reductase) in erythrocytes. *J. Lab. clin. Med.* **72**: 339, 1968.
31. SCOTT, E. M. A comparison of two methods of determining DPNH-methemoglobin reductase. *Clin. chim. Acta* **23**: 49, 1969.
32. BEUTLER, E. and BALUDA, M. C. A simple spot screening test for galactosemia. *J. Lab. clin. Med.* **68**: 137, 1966.
33. BEUTLER, E. A series of new screening procedures for pyruvate kinase deficiency, glucose-6-phosphate dehydrogenase deficiency and glutathione reductase deficiency. *Blood* **28**: 553, 1966.

ELECTROPHORETIC AND KINETIC CHARACTERIZATION OF A NADH-DIAPHORASE VARIANT IN A METHEMOGLOBINEMIC SUBJECT

JOEL M. SCHWARTZ, JONATHAN M. ROSS, PHILIP S. PARESS, KERRY FAGELMAN, and LAWRENCE FOGEL

Departments of Medicine, State University of New York, Downstate Medical Center and Coney Island Hospital affiliated with Maimonides Medical Center, Brooklyn, New York.

Hereditary methemoglobinemia which is not associated with an abnormality of globin has been shown by Scott to be due to a deficiency of a methemoglobin-reducing enzyme in the erythrocytes of affected individuals (1). This methemoglobin reductase enzyme specifically utilizes reduced nicotinamide adenine dinucleotide (NADH) as the electron donor. It may be assayed in hemolysate by the use of 2,6-dichlorobenzenone indophenol (DBI) dye which accepts electrons from the reduced enzyme at a rate 9,000 times more rapid than methemoglobin itself (2).

Deficiency of NADH diaphorase (methemoglobin reductase) has been established in at least 80 methemoglobinemic patients in various parts of the world (3). With few exceptions, it is unclear whether the deficiency in these patients is due to a defect in enzyme structure or to a diminished rate of enzyme synthesis. This communication gives data on a NADH diaphorase extracted from the red cells of a patient with hereditary methemoglobinemia, characterizing it as an enzyme variant with an abnormal structure and function.

The subject studied (4) was an intelligent 14-year-old girl whose blood methemoglobin level varied between 20 and 25% of the total hemoglobin. NADH-diaphorase activity could not be detected in her nitrited hemolysate (Fig. 1), and the washed nitrited red cells failed to reduce their methemoglobin when incubated in a medium containing glucose (Fig. 2). The capacity of the mother's hemolysate to reduce DBI and of her intact nitrited red cells to reduce methemoglobin was intermediate between that of the daughter and normal controls, suggesting that the mother is an acyanotic carrier of the methemoglobinemia gene.

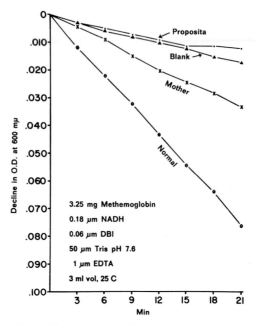

FIG. 1. Assay of NADH diaphorase in nitrited hemolysate (1). Normal enzyme transfers electrons from NADH to DBI, reducing the optical density at 600 mμ. Reaction in the patient's test cuvette lags behind reduction of DBI by NADH in the blank containing no hemolysate.

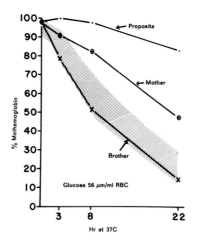

FIG. 2. Reduction of methemoglobin of washed nitrite-treated erythrocytes suspended in isotonic saline-phosphate buffer, pH 7.4, to which glucose is added (6). The shaded area indicates the variation obtained in experiments performed with erythrocytes from five normal subjects. The brother's hemolysate has normal NADH-diaphorase activity.

METHODS

Enzyme concentrates freed of hemoglobin were prepared according to suggestions of C. West (personal communication). Blood anticoagulated with heparin or acid-citrate-dextrose solution was centrifuged, plasma and buffy coat were removed, and the red blood cells were washed three times with 0.15 M potassium chloride buffered to pH 7.4. Five ml of washed red cells were hemolyzed with 20 ml of 0.005 M potassium phosphate buffer, pH 6.5, containing 0.1 mM disodium ethylenediamine tetraacetic acid (EDTA), and the stroma removed by centrifugation for 45 min at 27,000 × g. The clear supernatant fluid was dialyzed for $2\frac{1}{2}$ hr against two liters of 0.005 M potassium phosphate buffer, pH 6.5, containing 0.1 mM EDTA and 0.05 M potassium chloride, with the buffer changed once after $1^{1}/_{2}$ hr.

Dialyzed hemolysate was then applied to a 2.5 × 30 cm column of diethylaminoethyl (DEAE) Sephadex (Pharmacia A50), equilibrated with the same buffer. After the hemoglobin had been washed off the column with 125 ml of equilibrating buffer, the enzyme fraction was eluted with 0.005 M potassium phosphate buffer, pH 6.0, containing 0.1 mM EDTA and 0.3 M potassium chloride. The enzyme-rich eluate, measuring about 35 ml, was concentrated overnight by vacuum dialysis against 0.017 M Tris buffer, pH 7.55 containing 0.3 mM EDTA to a final volume of about 3 ml for the kinetic studies or 0.5 ml for the electrophoretic studies. The dialysis markedly reduced the concentration of potassium and chloride ions in the eluate. Solutions were made with double distilled water and all steps in the extraction process were performed in the cold.

NADH-diaphorase activity was determined at 25 C in 1 ml of solution containing 23.5 μmole Tris, 0.4 μmole EDTA, 0.06 μmole DBI, 0.06 μmole of NADH and 0.1 ml of enzyme concentrate. The final pH of the assay mixture was 7.55 and the ionicity was approximately 0.002. NADH was added to the test cuvette after the minor reduction of the dye initiated by the addition of the test sample had dissipated. The overall reaction was read at 600 mμ and was corrected for nonenzymatic reduction by subtracting the activity of a blank containing NADH but not enzyme concentrate. The enzyme concentrate was adjusted by dilution with dialysis solution to a concentration which provided a change in absorbance of 0.020 to 0.025/min when 0.1 ml of enzyme was used. In the studies of the activity of the enzyme, its thermal stability and the rate of utilization of NADH analogues, the reaction rate was calculated using the linear portion of the reaction curve between the second and sixth min after the addition of reduced pyridine nucleotide. A Gilford model 200 spectrophotometer linked to a multiple sample absorbance recorder was used to follow the reaction.

The Michaelis constant (K_m) for NADH was determined at 14 or 15 concentrations of NADH between 0.06 mM and 0.0027 mM, keeping the concentration of DBI at 0.06 mM. For the determination of K_m DBI, the NADH concentration was maintained at 0.06 mM and DBI concentration was varied between 0.12 mM and 0.012 mM. Eleven different DBI concentrations were used. In determining the DBI concentration in each cuvette at the start of the enzymatic reaction, account was taken of the nonenzymatic reduction of dye by the concentrate prior to the addition of NADH.

The conditions for the assay of NADH-diaphorase activity in nitrited hemolysates are given in Fig. 1 (1). The final concentration of DBI used in the assay of hemolysate is 0.02 mM and in the assay of enzyme concentrate is 0.06 mM. The reaction is first order with respect to DBI at both concentrations. Reduced nicotinamide adenine dinucleotide phosphate (NADPH) methemoglobin reductase was determined as NADPH-oxidase in the presence of methylene blue (5). *In vitro* reduction of methemoglobin by intact nitrited red cells was studied in the presence of glucose (6).

Freeze-thaw hemolysates and enzyme concentrates were subjected to horizontal starch gel electrophoresis at 4C in a Tris-EDTA-borate buffer system at pH 8.6 (7). A current of 10 ma for 16 hr or 30 ma for 4 hr was applied. The sliced gels were stained for NADH diaphorase and NADPH-methemoglobin reductase according to the tetrazolium method of Kaplan and Beutler (8). The staining reaction for NADH diaphorase is shown in Fig. 3.

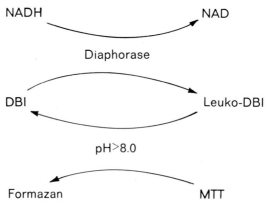

FIG. 3. Sequence of reactions in the detection of NADH diaphorase after starch gel electrophoresis (8). The buffered reaction mixture contains NADH, DBI and tetrazolium salt.

RESULTS

Upon electrophoresis of stroma-free hemolysate on starch gel at pH 8.6, normal NADH diaphorase migrates as a single band anodal to hemoglobin A_2. Fig. 4 shows that the normal band is absent from our patient's hemolysate. However, a faint band is observed with mobility intermediate between normal diaphorase and the rapidly migrating California variant described by Kaplan and Beutler (8). The parents' hemolysates each demonstrate a very faint band corresponding to the daughter's diaphorase variant as well as a band of normal mobility. The mother's pattern is shown in Fig. 4, position 5. The presence of the abnormal isoenzyme in the hemolysates of the parents constitutes proof of the heterozygous state.

The enzyme extraction process results in a 50- to 100-fold concentration of NADH-diaphorase activity per mg protein with a 50 to 90% yield. (For the comparison of the enzymatic activity in hemolysate and concentrate, both were assayed at a final DBI concentration of 0.02 mM.) The diaphorase activity with NADPH as substrate is reduced by about two-thirds. The NADH- and NADPH-methemoglobin reductase enzymes in the extract were each made visible after electrophoresis by overlaying the sliced gels with staining mixtures. When NADH was used in the staining solution, only NADH methemoglobin reductase was visualized. NADH-methemoglobin reductase and, in much lesser intensity, NADPH-methemoglobin reductase were visualized when NADPH replaced NADH in the staining solution (Fig. 5).

The activity of the NADH-diaphorase variant was estimated by comparing the specific activities of the enzyme concentrates extracted from methemoglobinemic and normal blood. The activity of the NADH-diaphorase variant was about 7% of normal: mean $\Delta OD/min$ per mg protein in the methemoglobinemic subject (five enzyme concentrates) = 0.10, and in seven normal subjects (14 enzyme concentrates) = 1.50.

Table 1 compares rates of utilization by variant and normal enzyme concentrates of NADH, deamino NADH and NADPH as substrates in the diaphorase reaction. The diaphorase variant does not utilize deamino NADH at the expected rate. Although NADPH is utilized proportionally at a higher rate by variant concentrate than by normal concentrate, there is no difference in the specific activity of the concentrates under these conditions: mean $\Delta OD/min$ per mg protein in the methemoglobinemic subject (three enzyme concentrates) = 0.07, and in four normal subjects (seven enzyme concentrates) = 0.07. Inasmuch as the

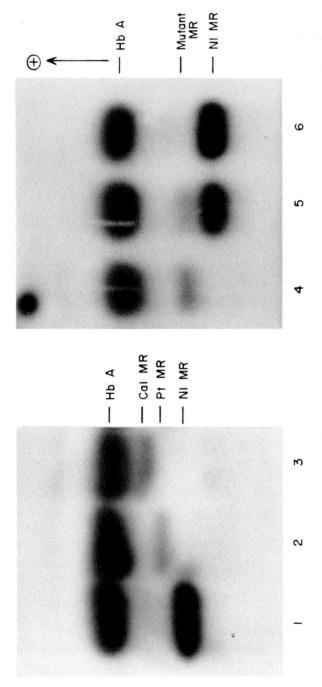

FIG. 4. Horizontal starch gel electrophoresis of NADH-methemoglobin reductase (MR) at pH 8.6. Two hundred μl of 1:2 freeze-thaw hemolysate were added to each slot. 1) Normal enzyme, 2) patient's enzyme, 3) California variant, 4) patient's enzyme, 5) mother's enzyme pattern, 6) normal enzyme.

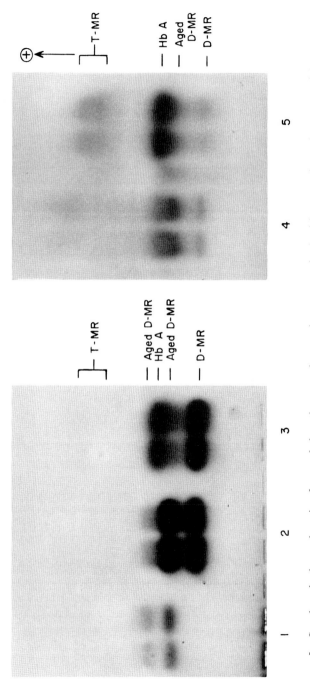

FIG. 5. Starch gel electrophoresis of normal hemolysate and normal extracts stained with a reaction mixture containing NADH [1, 2, 3] or NADPH [4, 5]. The locations of the NADH- and NADPH-methemoglobin reductases, as shown, have been established by Kaplan and Beutler in electrophoretic studies of hemolysates of normal individuals and individuals deficient in the respective enzymatic activities (8), and by J. C. Kaplan (personal communication). Weak cross-activity is observed for each of the enzymes in the hemolysate [3, 5]. Electrophoresis of the extracts reveals "aging" of NADH-methemoglobin reductase (13) with the development of new bands anodal to the original zone of activity [1,2,4]. Only NADH-methemoglobin reductase stains when the reaction mixture contains NADH [2]. NADH-methemoglobin reductase and, in lesser degree, NADPH-methemoglobin reductase stain when the reaction mixture contains NADPH, [4]. D-MR and T-MR are NADH- and NADPH-methemoglobin reductase. 1) Normal extract after one month cold storage, 2, 4) normal extract, 3, 5) normal hemolysate.

TABLE 1. *Relative utilization of NADH analogues*

Type of subject	Number of subjects	Relative deamino NADH utilization	Number of subjects	Relative NADPH utilization
Normal	5 (9)	96%	4 (8)	5%
Methemoglobinemia	1 (4)	69%	1 (3)	75%

Figures in parentheses indicate the number of enzyme concentrates tested. Rates of utilization of analogues are expressed as a percent of the reaction rate using NADH. Final concentration of NADH or its analogues in the assay mixture is 0.06 mM. Utilization rates are the average of duplicate determinations and were measured in a linear portion of the reaction curve between the second and sixth min.

TABLE 2. K_m *for NADH and DBI*

Type of subject	Number of subjects	K_m NADH μM Avg.	K_m NADH μM Range	Number of subjects	K_m DBI μM Avg.	K_m DBI μM Range
Normal	5 (7)	1.0	0.5 to 2.1	5 (6)	152	116 to 191
Methemoglobinemia	1 (3)	4.1	3.6 to 4.6	1 (3)	28	23 to 34

Figures in the parentheses indicate the number of enzyme concentrates tested. Initial reaction velocities were determined in duplicate at multiple concentrations of NADH or DBI as described in the text. K_m values for NADH were calculated from the regression of S/V against S, and for DBI from the regression of V against V/S. S is substrate concentration; V is initial velocity.

patient's erythrocyte NADPH-methemoglobin-reductase activity is normal, this finding suggests that the variant NADH-methemoglobin-reductase enzyme retains normal cross reactivity with NADPH.

The K_m for NADH and DBI were determined in five normal subjects and in the patient with hereditary methemoglobinemia (Table 2). The K_m for NADH of the diaphorase variant is higher than normal and the K_m for DBI of the diaphorase variant is lower than normal.

The stability of variant and normal enzyme concentrates was tested in incubation experiments at several temperatures (Table 3). The variant NADH diaphorase decayed more rapidly than normal diaphorase after incubation for 20 min at 38, 40 and 42 C. The same trend was apparent in longer experiments conducted at 38 and 40 C. The NADH-diaphorase variant lost about one half of its activity in the course of 1 hr, most of the loss occurring in the first 20 min.

TABLE 3. *Thermostability of diaphorase in normal and variant enzyme concentrates*

	Normal (%)	Variant (%)
Preincubation activity	100	100
Activity after 20 min		
at 38 C	89	64
40 C	88	56
42 C	83	56
Incubation at 38 C		
0 time	100	100
20 min	96	68
50 min	90	58
70 min	86	48
Incubation at 40 C		
0 time	100	100
20 min	92	60
40 min	84	55
60 min	79	50

Preincubation activity of freshly prepared enzyme concentrates were adjusted to ΔOD/min per 0.1 ml sample of 0.02 to 0.025 by the addition of dialysis solution. The protein concentration of the diluted normal enzyme concentrate was increased to that of the variant concentrate by the addition of salt-free bovine serum albumin. Aliquots (230 µl) of concentrate were distributed into 10×75 mm siliconized glass tubes and incubated in water baths at the temperatures shown. Activity of each aliquot was determined in duplicate and the residual activity of the incubated samples was expressed as a percent of the activity at zero time.

DISCUSSION

Determinations of NADH diaphorase performed with enzyme concentrate are more sensitive, accurate and specific for methemoglobin reductase than diaphorase determinations performed with hemolysate. Some of the reasons for these advantages are outlined in Table 4. The sensitivity of the diaphorase assay is improved by concentrating the enzyme activity 80-fold on the average, and by increasing the final concentration of DBI in the assay mixture from 0.02 to 0.06 mM. (The reaction is first order with respect to DBI at the concentrations used.) This threefold increase in the concentration of DBI can be accomplished without loss of accuracy because of the removal of methemoglobin from the sample, obviating nonenzymatic flow of electrons from reduced DBI to methemoglobin as the diaphorase reaction proceeds.*

* The regeneration of oxidized DBI when diaphorase is assayed in nitrited hemoylsate explains the smaller change in absorbance in the patient's test cuvette than in the no-hemolysate blank which is seen in Fig. 1. Increasing the DBI concentration exaggerates the oxidation of reduced dye by methemoglobin in hemolysate.

TABLE 4. *Several properties of concentrate compared to hemolysate, and their significance for the diaphorase determination*

50- to 100-fold increase in NADH-diaphorase activity per mg protein

Methemoglobin is removed
    No regeneration of DBI by leuko-DBI^{e-} → methemoglobin
    Higher DBI concentration can be used

NADH, NADPH, reduced glutathione, ascorbate, and ferrous Hb are removed
    Limits nonenzymatic reduction of DBI

Oxidized glutathione and pyruvate are removed
    Prevents generation of reduced glutathione via glutathione reductase
    Prevents oxidation of added NADH via lactate dehydrogenase

---

Although enzymes which remain in the concentrate are capable of utilizing NADH (or NADPH) as cofactor, there is no significant NADH (or NADPH) oxidase activity in the concentrate apart from diaphorase activity; cuvettes containing all reactants except dye show no decline in optical density during 10 min of observation at a wavelength of 340 mµ. This is not always the case for hemolysate. When diaphorase is assayed in crude hemolysate, reduced glutathione may be generated enyzmatically after the addition of NADH (9). The reduced glutathione directly reduces DBI, and spuriously elevates apparent NADH-methemoglobin-reductase activity.

The demonstration of altered electrophoretic mobility of the NADH-diaphorase derived from our patient's erythrocytes indicates that this enzyme is structurally abnormal. The structural change is associated with altered affinity for NADH and DBI and a lower rate of utilization of deamino NADH.

At least two characteristics of the variant enzyme may be germane to its function *in vivo*. The concentration of NADH in red blood cells is about 4 µM (10). Because of reduced affinity for NADH, the variant diaphorase is capable of achieving only half maximal velocity at this concentration of substrate. Although the functional properties of NADH-methemoglobin reductase might well vary with different electron acceptors, the demonstration of a high K_m for NADH in an assay system using DBI suggests that important low affinity for NADH may also occur when methemoglobin is reduced. Another characteristic of variant diaphorase which may be relevant to its intracellular function is the moderately increased thermolability *in vitro* at 38 C. The possibility that the variant diaphorase decays at an excessively rapid rate during the life span of the red cell is currently under investigation.

Several different electrophoretic phenotypes of NADH diaphorase have been recognized in patients with hereditary methemoglobinemia. At alkaline pH, the NADH-diaphorase variant described by West et al. (11) carries a more positive charge and the California variant (8) a more negative charge than our patient's diaphorase variant. The relationship of these three diaphorase phenotypes to two others characterized electrophoretically at neutral pH by Bloom and Zarkowski (12) is not yet known. Not all NADH-diaphorase variants are functionally abnormal. Detter et al. (13) have recently found a fast diaphorase isoenzyme present in heterozygous combination with normal diaphorase in the erythrocytes of an acyanotic individual. This isoenzyme has thus far been distinguished from normal diaphorase only by the difference in charge. Its presence is not accompanied by deficient NADH-diaphorase activity or impaired reduction of methemoglobin *in vitro*. Jaffé observed a similar phenomenon in an Italian family (14).

In addition to the principal NADH-diaphorase enzyme of human erythrocytes, Scott (9, 15, 16) has described a minor component which accounts for approximately 10% of the total NADH-diaphorase activity of normal blood cells and for the residual erythrocyte diaphorase activity in Alaskan patients with hereditary methemoglobinemia. The two diaphorase enzymes, also called NADH-dehydrogenases I and II, are distinguished by their chromatographic behavior and by a number of functional properties. The NADH-diaphorase variant described in this communication probably differs from Scott's NADH-dehydrogenase II in that it retains considerable reactivity with deamino NADH and it is demonstrated (by electrophoretic techniques) in methemoglobinemic blood, but not in normal blood.

I wish to acknowledge the statistical analysis provided by Dr. Stuart Kahan of the Division of Computer Science, State University of New York Downstate Medical Center, supported in part by Grant FR00291 from the U.S. Public Health Service.

Supported in part by United States Public Health Service General Research Support Grant FR05497 from the National Institutes of Health, Bethesda, Maryland.

REFERENCES

1. SCOTT, E. M. The relation of diaphorase of human erythrocytes to inheritance of methemoglobinemia. *J. clin. Invest.* **39**: 1176, 1960.
2. SCOTT, E. M. and MCGRAW, J. C. Purification and properties of diphosphopyridine nucleotide diaphorase of human erythrocytes. *J. biol. Chem.* **237**: 249, 1962.
3. JAFFE, E. R. DPNH methemoglobin reductase, in: Yunis, J. J. (Ed.), "Biochemical methods in red cell genetics." New York, Academic Press, 1969, p. 231.
4. SCHWARTZ, J. M. and GOLDMAN, B. Hereditary methemoglobinemia in a Puerto Rican family. *Clin. Res.* **15**: 287, 1967.

5. HUEHNNEKENS, F. M., CAFFREY, R. W., BASFORD, R. E. and GABRIO, B. W. Erythrocyte metabolism. IV. Isolation and properties of methemoglobin reductase. *J. biol. Chem.* **227**: 261, 1957.
6. JAFFE, E. R. The reduction of methemoglobin in erythrocytes of a patient with congenital methemoglobinemia, subjects with erythrocyte glucose-6-phosphate dehydrogenase deficiency, and normal individuals. *Blood* **21**: 561, 1963.
7. HUEHNS, E. R. and SHOOTER, E. M. Human hemoglobins. *J. Med. Genet.* **2**: 48, 1965.
8. KAPLAN, J. C. and BEUTLER, E. Electrophoresis of red cell NADH- and NADPH-diaphorase in normal subjects and patients with congenital methemoglobinemia. *Biochem. biophys. Res. Commun.* **29**: 605, 1967.
9. SCOTT, E. M. Congenital methemoglobinemia due to DPNH-diaphorase deficiency, in: Beutler, E. (Ed.), "Hereditary disorders of erythrocyte metabolism." New York, Grune and Stratton, 1968, p. 102.
10. LODER, P. B. and DEGRUCHY, G. C. Red cell enzyme and co-enzymes in non-spherocytic congenital hemolytic anemias. *Brit. J. Haemat.* **11**: 21, 1965.
11. WEST, C. A., GOMPERTS, B. D., HUEHNS, E. R., KESSEL, I. and ASHBY, J. R. Demonstration of an enzyme variant in a case of congenital methaemoglobinaemia. *Brit. med. J.* **4**: 212, 1967.
12. BLOOM, G. E. and ZARKOWSKI, H. S. Heterogeneity of the enzymatic defect in congenital methemoglobinemia. *New Engl. J. Med.* **281**: 919, 1969.
13. DETTER, J. C., ANDERSON, J. E. and GIBLETT, E. R. NADH diaphorase: An inherited variant associated with normal methemoglobin reduction. *Amer. J. hum. Genet.* **22**: 100, 1970.
14. JAFFE, E. R. Is it bad to be blue? *New Engl. J. Med.* **281**: 957, 1969.
15. SCOTT, E. M. Purification of diphosphopyridine nucleotide diaphorase from methemoglobinemic erythrocytes. *Biochem. biophys. Res. Commun.* **9**: 59, 1962.
16. SCOTT, E. M., DUNCAN, I. W. and EKSTRAND, V. The reduced pyridine nucleotide dehydrogenases of human erythrocytes. *J. biol. Chem.* **240**: 481, 1965.

DISCUSSION

S. W. Moses (*Israel*): We know from experience that very small infants do show a tendency to methemoglobinemia. Could it possibly be related to your findings of the absence of certain NADH or NADPH diaphorases in their erythrocytes? Do you always find this absence or do you find it only in certain cases? I am asking this question, because some infants show a greater tendency to develop methemoglobinemia during infection. Because acquired methemoglobinemias are rare in infants over the age of three months, it would be most interesting to get some information on the pattern of the different bands as a function of development.

E. Hegesh (*Israel*): The low level of NADH-diaphorase activity found in cord blood and in the blood of newborns and young infants by the methods accepted to date, could not be demonstrated electrophoretically. The absence of the NADPH 5 diaphorase band in adult blood can, in our view, not explain the enzymic deficiency in cord blood. However the findings may be due to technical reasons. In our technique, equal amounts of cord and adult blood extracts were used and staining was continued until all bands could be made visible. Therefore the strength of coloration of the different bands may not be proportional to the activity of the enzymes responsible for each band.

E. Kleihauer (*West Germany*): I am not convinced that only diaphorase deficiency is responsible for the high susceptibility of red cells of newborn infants to form methemoglobin. I think it's more the higher oxidation rate of fetal hemoglobin. In addition, comparative studies have shown that diaphorase activity is very similar in a young red cell population of cord blood and in that of adult blood, while it is significantly lower in a mixed and old red cell population of cord blood as compared to adult blood.

E. Hegesh: As far as it is known to me, it has not been established whether or not the low NADH-diaphorase activity in the red blood cells of newborns and young infants is responsible for the high levels of methemoglobin which were described by some investigators. The low enzymic activity could certainly be a contributing factor shifting the equilibrium of methemoglobin-hemoglobin to the oxidized form of the pigment.

J. C. Kaplan (*France*): I would like to point out the fact that you found a special pattern for the NADPH diaphorase in cord blood red cells. Is it a mere coincidence with the low NADH-diaphorase activity which is found in these cells? My second question concerns your control experiments. Among seven controls where you omitted several constituents of the reaction mixture in the fourth tube, there were still two bands. What was eliminated from this tube?

E. HEGESH: In the component study of the staining mixture prepared by your method, the component which was omitted in column four was 2,6-dichlorophenol-indophenol. It was clearly shown that in the absence of this dye, formazan formation proceeds and a specific staining for diaphorase can be achieved.

The properties of your mutant enzyme were investigated using a crude extract which was, more or less, free of hemoglobin. Thus the preparation included a lot of different enzymes, especially all NADH and NADPH diaphorases. I think that K_m studies, heat stability tests and studies of enzyme properties, in general, should be carried out on relatively pure proteins.

J. M. SCHWARTZ (*USA*): I don't know that it is necessary to work with a pure enzyme preparation. But it is necessary to separate the enzyme under investigation from substances capable of significantly affecting the kinetic measurements which are made. The extract used in these studies was freed of hemoglobin and of small molecules able to influence the NADH-diaphorase reaction (see Table 4, p. 144). The NADH-diaphorase activity averaged an 80-fold increase compared to hemolysate. The diaphorase activity with NADPH was reduced by about two-thirds, and it was shown by electrophoresis of variant and normal extracts concentrated to small volume that nearly all of this activity was in fact cross-acting NADH methemoglobin reductase. The NADPH-methemoglobin reductase contaminating variant and normal extracts were present in only trace amount. They could not be seen when stained with a reaction mixture containing NADH (see Fig. 5, p. 141).

SESSION IV

Chairmen: D. Busch, *West Germany*
S. Moses, *Israel*

Participants: I. Ben-Bassat, *Israel*
G. Bianco, *Italy*
F. Brok-Simoni, *Israel*
L. Eylon, *Israel*
E. Gallo, *Italy*
A. Hershko, *Israel*
Ch. Hershko, *Israel*
G. Izak, *Israel*
A. Karsai, *Israel*
J. Mager, *Israel*
U. Mazza, *Italy*
G. P. Pescarmona, *Italy*
B. Ramot, *Israel*
G. Ricco, *Italy*
R. Schwartz, *Israel*

SOME BIOCHEMICAL PARAMETERS OF DIFFERENTIATION AND MATURATION OF SYNCHRONOUS ERYTHROID CELL POPULATIONS

CH. HERSHKO, A. KARSAI, L. EYLON, R. SCHWARTZ and G. IZAK

Hematology Research Laboratory, Hadassah University Hospital and Hebrew University–Hadassah Medical School, Jerusalem, Israel

The process of cell maturation has long been a puzzling problem. Various schemes have been suggested, on the basis of kinetic studies, to explain the dynamics of cell renewal as it occurs in hemopoietic tissue. It is generally accepted (1) that there exists a stem cell compartment which multiplies continuously at the desired rate. Part of its derivatives remain stem cells, while others differentiate into one of the morphologically recognizable precursor compartments. Through a series of mitoses and the process termed maturation they reach a stage when they leave the bone marrow for their assigned functions in the periphery. Despite the work invested, our understanding of the basic processes occurring in these cells through the various stages just mentioned and the mechanism regulating the constant cell renewal is incomplete. One reason has been the heterogeneity of bone marrow populations, precluding meaningful interpretation of the data obtained in a pool of mixed cell types. If one could study each of the different cell types that populate the bone marrow separately, it might be possible to give more meaning to the terms "differentiation" and "maturation" than that offered by a morphological examination alone.

By utilizing the selective toxicity of actinomycin D we have obtained reasonably homogeneous erythroid populations at various stages of maturation (2). Changes in some biochemical parameters inherent in the differentiation and maturation of various cell populations thus produced are reported.

MATERIALS AND METHODS

Twenty rabbits weighing 2,500 to 3,500 g were bled daily for six consecutive days to a total of 200 ml, which slightly exceeds their calculated blood volume.

Hemoglobin (cyanmethemoglobin), hematocrit and reticulocyte counts were determined before and after the bleeding procedure, as well as on the day of the experiment and serially thereafter. On the last day of bleeding, 100 mg/kg of actinomycin D (supplied through the courtesy of Merck, Sharp and Dohme (Westpoint, Pa.) was injected s.c. Vials of 0.5 mg were dissolved in sterile saline solution shortly before use. Paired animals were sacrificed by exsanguination at varied intervals after administration of the drug.

A second group of 12 animals received an identical dosage of actinomycin D, but without the bleeding procedure described above. These animals were similarly sacrificed by exsanguination at various intervals after treatment. Bone marrow from both humeri, tibiae and femora was collected into cold buffered Krebs-Ringer solution. The pooled bone marrow was gently dispersed with a glass rod, and passed repeatedly through a 26 gauge needle to obtain a homogeneous cell suspension. The cell suspensions were washed three times in cold Krebs-Ringer solution, and resuspended in exactly 12 ml of the medium. The number of nucleated cells in the suspension was determined in a counting chamber. Smears made from the suspension were stained with supravital and May-Grünwald-Giemsa stains for viability and differential counts respectively (3). The suspensions were adjusted to contain 1.5 to 2.0 × 10^5 nucleated cells/mm^3. The following determinations were made on the cell suspensions before and 3 hr after incubation: Hb (cyanmethemoglobin), DNA (4), RNA (5), ATP (6) and heme (7).

Aliquots of 0.2 ml of the cell suspension were brought to a final volume of 1 ml by adding 0.4 ml pooled normal rabbit serum, 0.3 ml buffered Krebs-Ringer solution and 0.1 ml of an 0.4% glucose solution containing 1,000 units of penicillin and one of the following labeled precursors: 0.1 to 0.14 μc $Fe^{59}SO_4$ (supplied through the courtesy of Abbott Laboratories, Chicago, Ill.), specific activity 20 mc/mg Fe, 0.5 μc thymidine 6-T, 0.5 μc uridine 5-T and 0.2 μc 2-C^{14} glycine (supplied through the courtesy of Radiochemical Centre, Amersham, U.K.). The final mixtures were incubated for 1 and 3 hr. Incubation was terminated by washing the cells three times in cold buffered saline. The radio-iron incorporation into the heme moiety, isolated according to Thunnell (7), was estimated using a well type scintillation counter (Elron, Haifa, Model Nis, 17-P). The incorporation of glycine into protein was determined by measuring the radioactivity in trichloroacetic acid precipitate extracted repeatedly with 1:3 alcohol-ether. Incorporation of glycine into heme was measured in hemin isolated as above. 2-C^{14}-glycine incorporation was determined using a gas-flow thin-window β-counter at infinite thickness (Nuclear Chicago, Model 181B). DNA and RNA were separated from the cells as previously described (8) and the thymidine and uridine uptakes were determined in the isolated fractions using a Packard Model 3375 Tri-carb liquid scintillation spectrometer. The results were expressed per hr in μmole of substrate incorporated, or count/min per 10^9 nucleated cells. In some instances, 12 ml aliquot of the cell suspensions were cultured for 24 hr in a medium made up of four parts of pooled normal rabbit serum and six parts of Geys solution (9) containing 100 μg/ml streptomycin and 10,000 units/ml penicillin. The suspensions were adjusted to contain 2 to 3 × 10^3 nucleated cells/mm^3 and after equilibration with an O_2-CO_2 mixture (95%, 5% respectively) they were cultured in suitable sealed containers for 24 hr at 37 C. The incorporation activity of the cells following culture was determined as described above. Siliconized glassware was used throughout.

RESULTS

Peripheral blood. Hb levels and reticulocyte counts of the two experimental groups are presented in Table 1. In the normal group, Hb levels remained fairly constant throughout the experiment, starting with a mean value of 13.5 g/100 ml on the day of actinomycin D injection, with a slight drop to 11.1 g/100 ml on day 9. The reticulocyte count dropped from 4.5% to 0 on days 3 and 5, and was still very low on day 9. In the second group of animals, the bleeding procedure brought about a reduction of mean Hb from 12.8 to 6.3 g/100 ml, and a reticulocytosis of 29%. Throughout the nine-day period following the administration of actimonycin D, the mean Hb level remained within the range of 5.6 to 6.7 g/100 ml. As in the group of normal animals, the reticulocyte count dropped to 0 on the fifth day of the experiment, but rose again to 17.7% on the ninth day of the experiment.

Bone marrow cellularity and morphology. Since bone marrow was collected from the same bones under similar conditions, the total number of cells obtained served as a gross indicator of the cellularity of the hemopoietic tissue (Fig. 1). In the group of normal animals actinomycin D brought about a moderate reduction in bone marrow tissue, followed by a gradual recovery which did not, however, reach the initial values nine days after administration of the drug. The changes in marrow cellularity in the group of bled animals were more pronounced. Here, actinomycin administration caused a drastic reduction in the number of cells, followed by rapid recovery with a return to the initial values after nine days. It is evident from Fig. 1 that, while the two groups are comparable at the lowest point, the bled

TABLE 1. *Hemoglobin and reticulocyte counts in normal and bled rabbits*

	Time		Initial	1st day	3rd day	5th day	7th day	9th day
Hb g/100 ml	Normal (12 rabbits)	Mean	13.5	12.9	13.3	12.0	10.7	11.1
		Range	12.7 to 14.3	11.2 to 14.5	13.0 to 13.6	9.7 to 14.9	9.0 to 12.5	11.1 to 11.1
	Bled (20 rabbits)	Mean	6.3	5.6	5.7	6.7	6.0	6.5
		Range	5.5 to 7.7	4.9 to 6.3	5.5 to 6.0	6.6 to 6.8	5.0 to 6.7	6.4 to 6.5
Reticulocytes %	Normal (12 rabbits)	Mean	4.5	0.3	0	0	0.2	0.15
		Range	3.8 to 5.3	0 to 0.9			0 to 0.4	0.1 to 0.2
	Bled (20 rabbits)	Mean	29.0	31.0	2.4	0	5.4	17.7
		Range	22.0 to 42.0	19.0 to 43.0	1.4 to 3.3		3.0 to 8.0	14.8 to 20.6

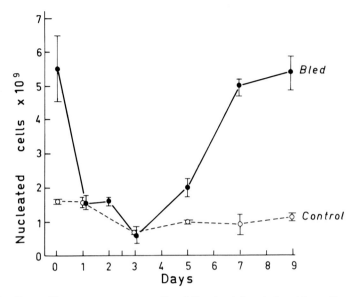

FIG. 1. Rate of bone marrow regeneration following injury induced by actinomycin, in bled and unbled rabbits. (Vertical lines denote the limits of SE.)

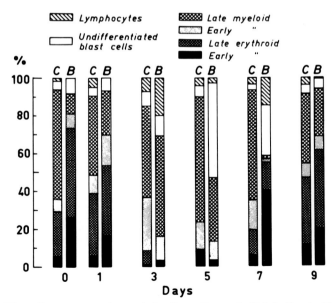

FIG. 2. The cell types at varying periods after actinomycin D injection in bled (B) and unbled (C) rabbits.

TABLE 2. Nucleic acid synthesis and cellular RNA content

Source of marrow		Number of procedures[b]	Count/min x 10³ per hr/10⁹ nucleated cells				RNA content µg/10⁹ nucleated cells	
			DNA		RNA			
			Mean	Range	Mean	Range	Mean	Range
Nonbled controls	Before culture	8	1.1	0.7 to 1.63	0.9	0.74 to 1.3	4.8	3.7 to 6.4
	After culture	8	1.4	1.1 to 2.4	1.6	1.2 to 2.4		
Bled controls	Before culture	8	21	18.5 to 24.3	3.4	3.1 to 4.4	3.7	3.1 to 5.8
	After culture	8	12	9.6 to 13.9	6.1	5.2 to 8.1		
5 days[a]	Before culture	8	93	87.0 to 98.2	71	64.0 to 76.5	17.0	14.6 to 19.2
	After culture	8	34	29.8 to 38.3	22	18.4 to 26.3		
7 days[a]	Before culture	12	14	11.8 to 17.1	19	16.6 to 21.8	5.1	4.3 to 6.9
	After culture	12	6	4.9 to 7.6	18.5	17.1 to 22.4		
9 days[a]	Before culture	8	1.0	0.83 to 1.6	1.1	0.85 to 1.35	5.4	4.8 to 6.9
	After culture	8	1.8	1.4 to 2.3	3.6	3.3 to 4.2		

[a] After actinomycin.
[b] Each rabbit provided two duplicates for 1 and 3 hr culture.

animals initially started out with hemopoietic tissue about four times more cellular than that of the normal controls, and that the regeneration process produced more than four times as many cells during the same period than in the bone marrow of the control animals.

The disparity in the cell types comprising the hemopoietic tissue in the two experimental groups of animals was even more pronounced than these quantitative changes (Fig. 2). In both the normal and the bled groups, the percentage of erythroid precursors had been greatly reduced on the third and fifth days after actinomycin D, injection, and completely replenished by the ninth day of the experiment. However, while the normal rabbit's marrow contained a wide variety of elements at each sampling following actinomycin, hemopoietic tissue of the bled animals consisted of fairly uniform population groups. Thus, one to two days following the administration of actinomycin D, the bulk of the cells were orthochromatic normoblasts. Three days later, large undifferentiated mononuclear cells replaced the late normoblasts, while the picture on the ninth day after actinomycin reverted to that seen on day zero.

Synthetic activity. The following data are related exclusively to the group of bled animals.

Nucleic acid synthesis: The replication rate of DNA was most rapid in the undifferentiated cell population five days following actinomycin administration (Table 2). It diminished steadily parallel with the emergence of erythroid precursor cells, until on the ninth day of the experiment DNA synthesis was comparable with the rate found in the normal control animals. Following 24 hr incubation of the marrow cells, DNA synthesis diminished markedly at all phases of the experiment.

The pattern of RNA synthesizing activity was similar to that of DNA, reaching its peak on the fifth day after administration of actinomycin D. It should be noted that the cellular RNA content rose substantially at the time of maximal RNA synthesis.

Heme synthesis: The rate of heme production, as judged by the incorporation of C^{14}-glycine into heme (Table 3), was highest in the population of late normoblasts isolated soon after actinomycin injection. In contrast with this cell population, the undifferentiated cells (day 5) produced very little heme, and this was also almost equally true with regard to the early erythroid cells (day 7). Only when the cells reached the poly- and orthochromatic stage, did their activity become similar to that seen initially. The culture of these suspensions for 24 hr produced cell populations which closely resembled those freshly isolated from the rabbit at the corresponding

TABLE 3. 2-C^{14}-glycine and Fe^{59} incorporation into heme

			μmole/10^9 nucleated cells/hr			
			2-C^{14}-glycine		Fe^{59}	
Source of marrow		Number of procedures[b]	Mean	Range	Mean	Range
Nonbled controls	Before culture	8	1.0	0.78 to 1.4	1.0	0.81 to 1.31
	After culture	8	0.75	0.64 to 0.84	0.88	0.59 to 1.1
Bled controls	Before culture	8	12.8	11.2 to 13.4	4.9	3.8 to 5.6
	After culture	8	14.1	12.8 to 15.6	2.6	1.4 to 3.3
1 day[a]	Before culture	8	17.3	16.1 to 19.0	11.7	10.3 to 13.1
	After culture	0	Not examined		Not examined	
2 days[a]	Before culture	8	10.3	8.7 to 12.1	7.3	6.4 to 8.2
	After culture	8	5.9	4.7 to 7.3	5.1	3.9 to 6.7
3 days[a]	Before culture	4	2.7	1.8 to 3.4	0.9	0.74 to 1.35
	After culture	0	Not examined		Not examined	
5 days[a]	Before culture	8	2.9	1.7 to 3.3	1.4	1.0 to 1.9
	After culture	8	3.6	3.0 to 4.2	0.8	0.71 to 0.98
7 days[a]	Before culture	12	3.1	2.4 to 4.1	4.6	3.4 to 6.1
	After culture	12	4.2	3.3 to 5.8	2.4	1.9 to 3.7
9 days[a]	Before culture	8	8.3	6.4 to 9.7	5.1	4.3 to 6.3
	After culture	8	2.7	1.6 to 3.5	3.3	2.1 to 4.8

[a] Days after actinomycin injection.
[b] Each rabbit provided two duplicates for 1 and 3 hr incubation.

intervals after actinomycin injection. Thus between days 1 and 3 while heme synthesis dropped from its maximal to its lowest rate, 24-hr incubation of the day 2 sample resulted in a similar trend in heme elaboration. The rate of heme synthesis, as studied by incorporation of Fe^{59}, (Table 3) was very similar to that found with C^{14}-glycine, with one notable exception: *in vitro* incubation resulted in an apparent reduction of heme synthesis in all samples examined.

Protein synthesis: Incorporation of C^{14}-glycine into protein followed a pattern that corresponded roughly with the rate of heme synthesis. Thus, the highest rates of protein synthesis were found on the first two days, and on days 7 to 9 of the experiment (Table 4).

TABLE 4. $2-C^{14}$-glycine incorporation into protein

Source of marrow		Number of procedures [b]	μmole/10^9 nucleated cells per hr	
			Mean	Range
Nonbled controls	Before culture	8	3.9	3.1 lo 4.3
	After culture	8	2.1	1.6 to 2.95
Bled controls	Before culture	8	18.6	15.6 to 20.4
	After culture	8	12.3	10.8 to 13.6
1 day [a]	Before culture	8	18.0	16.0 to 19.8
	After culture	0	Not examined	
2 days [a]	Before culture	8	15.1	13.9 to 17.4
	After culture	8	6.3	5.5 to 7.3
3 days [a]	Before culture	4	15.1	14.1 to 18.0
	After culture	0	Not examined	
5 days [a]	Before culture	8	14.3	12.4 to 16.3
	After culture	8	6.5	4.8 to 7.6
7 days [a]	Before culture	12	22.4	19.7 to 24.3
	After culture	12	24.3	21.8 to 26.2
9 days [a]	Before culture	8	25.1	23.1 to 27.0
	After culture	8	28.4	25.6 to 30.4

[a] Days after actinomycin injection.
[b] Each rabbit provided two duplicates for 1 and 3 hr incubation.

ATP content (Table 5): The cellular content of ATP was highest in the cell population obtained five days after actinomycin injection. The lowest values were observed in the bled animals prior to administration of the drug. Twenty-four hr culture of the cells resulted in a rise of cellular ATP content in all experiments.

DISCUSSION

Evidence presented here, as shown in other studies (10–12), points to an interference with stem cell differentiation as the site of action of actinomycin D on erythropoiesis *in vivo*. Thus, the relatively mild injury produced by this compound in the marrow of a normal rabbit compared to the effect of

TABLE 5. *Cellular ATP content*

Source of marrow		Number of procedures[b]	μmole $ATP/10^9$ nucleated cells	
			Mean	Range
Nonbled controls	Before culture	8	0.9	0.74 to 1.3
	After culture	8	1.2	1.0 to 1.76
Bled controls	Before culture	8	0.6	0.38 to 0.91
	After culture	0	Not examined	
5 days[a]	Before culture	8	3.1	2.7 to 4.2
	After culture	8	6.4	5.3 to 7.8
7 days[a]	Before culture	12	0.8	0.61 to 0.98
	After culture	12	1.6	1.2 to 2.4
9 days[a]	Before culture	8	1.1	0.85 to 1.6
	After culture	8	1.4	1.1 to 1.85

[a] Days after actinomycin injection.
[b] Each rabbit provided two duplicates for 1 and 3 hr incubation.

the drug on the hemopoietic tissues of bled animals may reflect a lower rate of erythroid oriented stem cell differentiation in the former compared with the latter group of rabbits. For the same reason normal bone marrow, in contrast to marrow stimulated by bleeding, failed to produce the synchronized groups of erythroid precursors so well demonstrated in those who had been bled.

The replication rate of DNA and RNA was highest in the undifferentiated blast cell population, present five days after the exposure of stimulated marrow cells to actinomycin D. With the emergence of differentiated erythroid precursor cells, and concomitant with their successive stages of morphological maturation, nucleic acid synthesis dropped steadily. This loss of synthetic activity has been demonstrated in two independent ways: 1) by comparison of the synthetic activity of rabbit marrow cells at increasing intervals after actinomycin injection and 2) by comparison of synthetic activity of the same marrow populations before and after 24 hr of culture *in vitro*. These findings are in accordance with the contention that loss of RNA-producing ability of the erythroid cells is inherent in their maturation

(13). It is conceivable that the reduction of nucleic acid synthesis during incubation can be attributed to circumstances unfavorable to further cellular development prevailing in the culture medium. However, the increase in cellular ATP content found in almost all cell populations following incubation militates against cellular damage due to inimical culture conditions.

While the above data may serve as biochemical indicators of cell proliferation, the study of heme synthesis, as measured by the incorporation of 2-C^{14}-glycine or Fe^{59} in heme, was designed to assess erythroid differentiation and maturation. It was found, using either tracer, that the rate of heme synthesis was very low in the earliest erythroid population observed during erythroid regeneration, increasing gradually thereafter to a peak rate about 24 hr before the emergence of orthochromatic normoblasts. As in the case of nucleic acid synthesis, these changes could be demonstrated both by comparing incorporation into bone marrow of animals sacrificed at different phases of erythroid regeneration, and by comparing the same samples before and after 24-hr culture. While this was the case using C^{14}-glycine, an unexpected apparent drop in heme synthesis was found after *in vitro* culture of the fifth and seventh day samples using Fe^{59}. This apparent difference might be the result of saturation of the incubated cells by an excess of "cold" iron abundantly present in the medium, which was not the case with glycine.

In contrast to the rapid loss of RNA synthesizing activity in the maturing erythroid precursors, protein synthesis (reflected by the incorporation of C^{14}-glycine) was still very high nine days after actinomycin injection. This finding tallies with evidence produced by others for the existence of a relatively stable messenger RNA in the maturing erythroid cells (14).

These observations on synthesis offer some correlation with the morphologic features of undifferentiated cells on the one hand, and the contrasting pattern of differentiating and progressively maturing cells on the other. In the present state of our knowledge we are unaware of the mechanisms responsible for this shift in synthetic activity brought about by differentiation and maturation. Do undifferentiated blast cells contain the apparatus for Hb formation, but lack one of the T fractions described recently by Lipmann (15)? Once these systems are fully operative, what is the mechanism which terminates Hb synthesis? Is it a negative feedback by the increasing amount of Hb within the cell? How and in what way is RNA broken down once the cells reach their appropriate stage of maturation?

It was with these thoughts in mind that we turned our early attention to the reticulocyte, as this cell can be produced, isolated and fractionated quite easily. In the course of these studies, which were performed mainly on rabbit and partly on human reticulocytes, we have learned some of the biochemical parameters of their maturation (16). When normoblasts lose their nuclei, the remaining reticulocytes are rich in RNA, the bulk of which is in particulate form comprising the polysomes which produce Hb. The polysomes dissociate into monosomes, which are then attacked by the ribonuclease present. From the data accumulated in our laboratory, it seems that this phenomenon is triggered by a loss of Mg^{++}, as Mg^{++}-chelating substances were found to enhance polysome disaggregation, while the addition of Mg^{++} markedly diminished the ribosomal RNA breakdown. It was also found that normal plasma and serum contain one or more dialyzable factors which are capable of triggering the disaggregation of polysomes. The oligonucleotides produced by ribonuclease action are further broken down until the free purines and pyrimidines leave the cell, while ribose-1-phosphate is reutilized in the pentose cycle.

At present we are unable to answer similar questions concerning earlier nucleated red cell precursors, although some progress has already been made. We were able to obtain ribosomes and soluble fractions from the synchronized red cell precursors, at various stages of maturation, produced by actinomycin. It may be assumed that the comparison of the synthetic activity of these subcellular fractions with those already studied in the reticulocytes (16), will provide further insight into the intricacies of the problems of erythroid precursor cells.

REFERENCES

1. STOHLMAN, F. Some aspects of erythrokinetics. *Seminars Hemat.* **4**: 304, 1967.
2. HERSHKO, CH., SCHWARTZ, R. and IZAK, G. Morphological-biochemical correlations in rabbit red cell precursors synchronized by actinomycin administration. *Brit. J. Haemat.* **17**: 569, 1969.
3. SCHWIND, J. L. The supravital method in the study of the cytology of blood and marrow cells. *Blood* **7**: 597, 1950.
4. CHARGAFF, E. and DAVIDSON, J. (Gen. Eds.) "Nucleic acid." New York, Academic Press, Inc., 1955, v. I, p. 285.
5. SCHNEIDER, W. C. Determination of nucleic acids in tissues by pentose analysis, in: Golowick S. P. and Kaplan, N. O. (Eds.), "Methods of Enzymology." New York, Academic Press, Inc., 1957, v. 3, p. 680.
6. STREHLER, B. L. and MCELROY, W. D. Assay of adenosine triphosphate, in:, Golowick, S. P. and Kaplan, N. O. (Eds.), "Methods in enzymology." New York, Academic Press, Inc., 1957, v. 3, p. 871.
7. THUNNELL, S. Determination of incorporation of Fe^{59} in hemin of peripheral red blood cells and of red cells in bone marrow culture. *Clin. chim. Acta* **11**: 321, 1965.
8. IZAK, G., KARSAI, A., EYLON, L. and SCHIFFMAN, A. Radiation-induced changes

in some biochemical parameters of the hemopoietic tissue of rabbits. *Israel J. med. Sci.* **6**: 14, 1970.
9. GEY, G. O. and GEY, M. K. The maintenance of human normal cells and tumor cells in continuous culture. *Amer. J. Cancer* **27**: 45, 1936.
10. REISSMANN, K. B. and ITO, K. Selective inhibition of erythroid cell differentiation in the mouse by low doses of Actinomycin D. *Ann. N.Y. Acad. Sci.* **149**: 193, 1968.
11. GURNEY, C. W. and HOFSTRA, D. Assessment of actinomycin and radiation damage of stem cells by the erythropoietin tolerance test. *Radiat. Res.* **19**: 599, 1963.
12. KEIGHLEY, G. and LOWY, P. H. Actinomycin and erythropoiesis and the production of erythropoietin in mice. *Blood* **27**: 637, 1966.
13. LAJTHA, L. G. Bone marrow cell metabolism. *Physiol. Rev.* **37**: 50, 1957.
14. KRANTZ, S. B. and GOLDWASSER, E. On the mechanism of erythropoietin-induced differentiation. 2. The effect on RNA synthesis. *Biochim. biophys. Acta (Amst.)* **103**: 325, 1965.
15. MULKIN, M. and LIPMANN, F. Fusidic acid; inhibition of factor T_2 in reticulocyte protein synthesis. *Science* **164**: 71, 1969.
16. IZAK, G., BEN-BASSAT, J., KARSAI, A. and MAGER, J. Biochemical changes in the maturing reticulocyte. Effects of ionizing radiations on the haematopoietic tissue. *Panel proceedings series. International Atomic Energy Agency, Vienna*, 1967, p. 25.

RED CELL METABOLISM IN IRON DEFICIENCY ANEMIA

U. MAZZA, G. P. PESCARMONA, G. BIANCO, G. RICCO and E. GALLO

Institutes of Medical Pathology and of Biological Chemistry, University of Turin, Turin, Italy

In 1968, Prato et al. (1) investigated some aspects of porphyrin metabolism in iron deficiency anemia (IDA). The results of our recent studies on the activity of some enzymes of the Embden-Meyerhof pathway and of hexosemonophosphate shunt, and new data on the synthesis of porphobilinogen (PBG) and porphyrin in iron deficient erythrocytes are now presented.

MATERIALS AND METHODS

The criteria for the diagnosis of IDA, and therefore for inclusion in the study, were as follows: Hb ≤ 8.5 g/100 ml, mean corpuscular Hb (MCH) ≤ 27 pg, mean corpuscular volume (MCV) ≤ 80 μ^3, serum iron concentration < 60 µg/100 ml, total iron binding capacity (TIBC) > 400 µg/100 ml

TABLE 1. *Hematological data in six patients with IDA*

Case no.	Sex	Age	Hb g/100 ml	RBC millions/mm^3	MCH	Hct %	MCV μ^3	Reticulocytes % RBC	Serum iron µg/100 ml [a]	TIBC µg/100 ml [b]	Cr^{51} $T_{\frac{1}{2}}$ days [c]	
1	M	56	8.2	4.05	21	31	77	15	60	420	26	Hemicolectomy
2	F	57	8.2	3.50	24	28	80	9	61	425	27	Gastrectomy
3	M	51	7.2	2.90	25	24	80	7	35	490	—	Intestinal carcinoma
4	F	56	7.0	2.80	25	22	78	11	45	460	—	Genital bleeding
5	F	45	6.9	2.80	25	21	75	6	45	510	—	Genital bleeding
6	F	21	5.8	3.04	20	23	76	16	50	540	22	Idiopathic hypochromic anemia

[a] Normal values: females 90 to 120 µg/100 ml; males 100 to 130 µg/100 ml.
[b] Normal values: 320 to 400 µg/100 ml. [c] Normal values: > 32 days.

and hypochromic, microcytic RBC on the blood smear. The IDA was due to various causes and responded to iron therapy (Table 1).

Hb was estimated spectrophotometrically as cyanomethemoglobin, and the hematocrit, RBC and reticulocyte counts were determined by standard methods. Serum iron was measured using potassium rhodanate after incineration by heating in sulphuric acid. The δ-aminolevulinic acid (ALA) dehydrase activity was determined by the slightly modified method of Gibson et al. (2) By this method the porphyrin formed from PBG could also be measured.

The activity of red cell enzymes was determined after centrifuging heparinized blood at $3,000 \times g$ for 5 min, discarding the plasma and the upper buffy layer and washing the RBC in 153 mM NaCl and 17 mM Tris buffer, pH 7.5. The cells were then suspended in a solution of 0.001 M Tris, 0.0015 M $MgCl_2$ and 0.03 M NaCl and hemolyzed by sonication for five seconds.

The activity of the following enzymes were determined: hexokinase (HK), phosphofructokinase (PFK) aldolase (ALD), phosphoglycerate kinase (PGK), glyceraldehydephosphate dehydrogenase (GAPDH), pyruvate kinase (PK), lactate dehydrogenase (LDH) and glucose-6-phosphate dehydrogenase (G6PD). Enzyme activity was assayed at 37 C according to the methods of Bucher et al. (3) with modifications in reactant composition (4).

Hb content was determined in the hemolysate and in whole blood; RBC counts were also done, so that it was possible to calculate the enzyme activities per ml of RBC, per g of Hb, and per 10^{10} cells.

RESULTS

The PBG and porphyrin synthesized from ALA were measured in eight female patients with IDA (5). The PBG synthesis in these patients was markedly increased. The actual amount of PBG formed was 60% higher than in normal controls (Fig. 1). By contrast, the conversion of the PBG, formed during the incubation, into porphyrin was found to be decreased by about 60% (Fig. 2).

The same experiments repeated during iron therapy showed further evidence for an increased ALA dehydrase activity (PBG + porphyrin synthesis) which was related to the number of reticulocytes. The porphyrin synthesis, reduced before therapy, increased gradually and was within normal limits when a complete hematologic recovery was achieved (Fig. 3).

The activity of glycolytic enzymes and of the G6PD was determined in six patients, whose hematological data are reported in Table 1. The enzyme

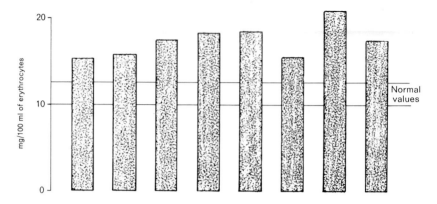

FIG. 1. Values of PBG synthesis from ALA (including the porphyrin formed from the PBG synthesized) in eight cases of IDA.

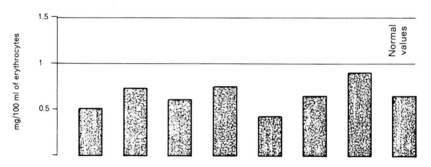

FIG. 2. Values of porphyrin synthesis from ALA in eight cases of IDA.

activities are summarized in Table 2 and Fig. 4. The mean and range of values are expressed as percent of normal activity. Similar results were obtained when the enzyme activity was expressed per g of Hb or per 10^{10} RBC. It is noteworthy that the values are scattered over a wide range. PFK activity was not significantly increased (9%); in only one patient was a 50% increase found. Mean activity of all other enzymes studied was raised (HK 75%, ALD 36%, GAPDH 91%, PGK 58%, PK 67%, LDH 116% and G6PD 73%). In all the determinations carried out before treatment, the increased enzyme activity did not seem to be related either to the degree of anemia or to the serum iron levels.

Two patients were followed up on iron therapy. A further increase in the activity of enzymes (except PFK) was evident during the reticulocyte

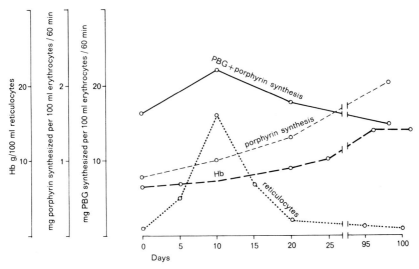

FIG. 3. Modification of PBG and porphyrin synthesis, Hb level and reticulocyte number induced by iron therapy in a case of IDA.

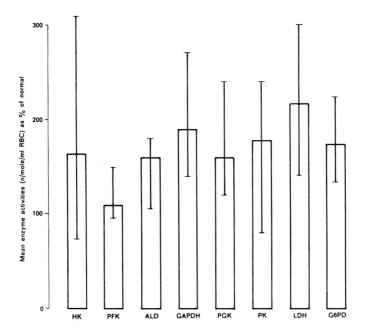

FIG. 4. Enzyme activity in six cases of IDA. Mean values and range are recorded as percent of normal activity in μmole/ml RBC.

TABLE 2. Activities of glycolytic enzymes and of G6PD activities

		HK	PFK	ALD	GAPDH	PGK	PK	LDH	G6PD
					Normal values [a]				
nmole/ml	RBC	20 ± 2.9	469 ± 40	117 ± 11	$3,023\pm648$	$4,171\pm678$	334 ± 29	$2,531\pm108$	$1,751\pm115$
nmole/g	Hb	58.5	1,371	342	8,839	12,195	976	7,400	5,078
nmole/10^{10}	RBC	17.9	419	104	2,703	3,729	292	2,263	1,481
					Patients affected by IDA				
1. nmole/ml	RBC	48	503	196	5,206	6,495	608	5,297	2,521
nmole/g	Hb	209	2,192	854	22,684	28,300	2,649	23,080	10,841
nmole/10^{10}	RBC	41	429	167	4,439	5,538	518	4,516	1,806
2. nmole/ml	RBC	34	522	134	4,718	5,819	582	4,785	3,109
nmole/g	Hb	139	2,131	547	19,257	23,751	2,375	19,530	12,436
nmole/10^{10}	RBC	32.5	499	128	4,510	5,562	556	4,574	2,926
3. nmole/ml	RBC	64	732	202	8,268	7,410	841	7,819	3,921
nmole/g	Hb	306	3,502	966	39,560	35,454	4,024	37,411	18,810
nmole/10^{10}	RBC	76	869	240	9,816	8,798	998	9,283	4,710
4. nmole/ml	RBC	13	426	119	4,227	4,940	261	3,571	—
nmole/g	Hb	36	1,177	329	11,677	13,646	721	9,865	—
nmole/10^{10}	RBC	12	401	105	3,941	4,430	215	3,004	—
5. nmole/ml	RBC	14	437	118	7,735	9,867	297	5,754	2,910
nmole/g	Hb	49	1,528	412	27,045	34,500	1,038	20,118	9,806
nmole/10^{10}	RBC	12	376	102	6,661	8,497	256	4,955	2,301
6. nmole/ml	RBC	39	463	177	4,600	5,048	766	5,523	2,415
nmole/g	Hb	178	2,114	808	20,994	23,050	3,498	25,219	10,810
nmole/10^{10}	RBC	34	403	154	4,010	4,390	666	4,803	2,186
Mean									
nmole/ml	RBC	35	514	158	5,792	6,596	559	5,458	2,975
nmole/g	Hb	153	2,107	653	23,534	26,450	2,381	22,537	12,540
nmole/10^{10}	RBC	39	515	158	5,887	6,557	599	5,626	2,786

[a] Mean ± 2 SD for 12 cases

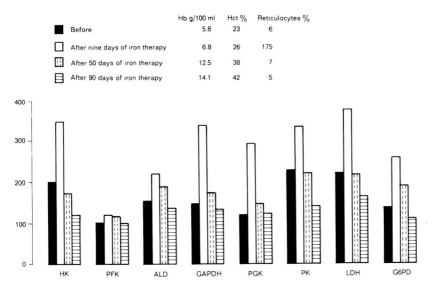

FIG. 5. Changes in enzyme activity, Hb, Hct and reticulocyte values induced by iron therapy in a case of IDA.

response. When clinical improvement ensued, the enzyme activities decreased and were almost normal upon attaining complete recovery (Fig. 5).

DISCUSSION

The ALA dehydrase activity was raised in all patients affected by IDA. Our results thus conform to those of Steiner, et al. (6) obtained in bone marrow aspirates. By contrast, Lichtman and Feldman (7) found normal values in hemolysates.

In all our cases tested, the amount of porphyrin synthesized from ALA was smaller than in normal controls, which correlates with the reduced activity of the enzymes involved in this reaction. Similar results were reported by Lichtman and Feldman (7).

The glycolytic enzyme activities were spread over a wide range, but generally they were significantly increased. Only the activity of PFK was not significantly altered. Our results therefore agree with those of Mac-Dougall (8), who studied the HK, LDH, G6PDH and 6PGDH activities in children with nutritional IDA, and those of Vuopio (9), Marks (10) and Tanaka (11) who investigated the activity of some enzymes of the Embden-Meyerhof pathway and of the hexosemonophosphate shunt.

The increased activity of these two groups of enzymes involved in apparently unrelated reactions probably depends on more than one cause. One notable factor, likely to be a determinant, is the presence of a higher number of young cells, related to the shortened erythrocyte life span observed in patients with IDA. Chromium survival studies carried out on patients 1,2 and 6, who did not have any apparent bleeding, showed a decreased RBC survival; the other patients (3, 4 and 5) had gastrointestinal or genital bleeding. That the younger RBC population is associated with increased enzyme activity is also supported by the fact that the activity further increased during the reticulocyte response.

The conversion of PBG to uroporphyrin has not yet been completely elucidated: at least two enzymes, namely PBG deaminase and uroporphyrin isomerase, are involved in this reaction. The reduced porphyrin synthesis observed in IDA seems to confirm the view that iron is directly involved in this reaction, as suggested by Lichtman and Feldman (7). We have observed that porphyrin synthesis gradually increases during iron therapy when the old RBC are replaced by newly formed RBC. These findings are in keeping with Lichtman's hypothesis that the iron does not act as a cofactor, but rather plays a role as an enzyme inductor in the erythroblasts at an earlier stage.

REFERENCES

1. PRATO, V., MAZZA, U., MASSARO, A. L., BIANCO, G. and BATTISTINI, V. Porphyrin synthesis and metabolism in iron deficiency anemia: II) *In vitro* studies. *Blut* **17**: 14, 1968.
2. GIBSON, K.D., NEUBERGER, A. and SCOTT, J. J. The purification and properties of delta-aminolevulinic acid dehydrase. *Biochem. J.* **61**: 618, 1955.
3. BUCHER, T., LUH, W. and PETTE, D. Einfache und zusammengesetzte optische Tests mit Pyridinnucleotiden, in: Lang, K. and Lehnartz, E. (Eds.), "Hoppe-Seyler Thierfelder: Handbuch der physiologisch- und patologisch-chemischen Analyse." Berlin, Springer Verlag, 1964, v. 6 part A, p. 292.
4. PESCARMONA, P. G., BOSIA, A. and ARESE, P. *Experientia* (*Basel*) (in press).
5. PRATO, V., MAZZA, U., MASSARO, A. L., BIANCO, G. and BATTISTINI, V. Porphyrin synthesis and metabolism in iron deficiency anemia: I) *In vivo* studies. *Blut* **16**: 333, 1967.
6. STEINER, M., BALDINI, M. and DAMASHEK, W. Enzymatic defects of heme synthesis in thalassemia. *Ann. N.Y. Acad. Sci.* **119**: 548, 1964.
7. LICHTMAN, H. C. and FELDMAN, F. *In vitro* porphyrin synthesis by iron deficient erythrocytes. *Proc. Soc. exp. Biol.* (*N.Y.*) **126**: 38, 1967.
8. MACDOUGALL, L. G. Red cell metabolism in iron deficiency anemia. *J. Pediat.* **72**: 303, 1968.
9. VUOPIO, P. Red cell enzymes in anemia. *Scand. J. clin. Lab. Invest.* Suppl. **72**: 1, 1963.
10. MARKS, P. A. Red cell glucose-6-phosphate, 6-phosphogluconate dehydrogenases and nucleoside phosphorylase. *Science* **127**: 1338, 1958.
11. TANAKA, K. R., VALENTINE, W. N. and MIWA, S. Pyruvate kinase deficiency hereditary nonspherocytic hemolytic anemia. *Blood* **19**: 267, 1962.

THE RED CELL NUCLEOSIDE MONOPHOSPHATE KINASES AND THEIR REGULATORY ROLE IN PURINE NUCLEOTIDE METABOLISM

A. HERSHKO, M. D. and J. MAGER, M. D., Ph. D.

Laboratory for Cellular Biochemistry, Department of Biochemistry,
Hebrew University–Hadassah Medical School, Jerusalem, Israel

Previous work (1) indicated that RBC are unable to carry out phosphorylation of IMP to IDP and ITP, while displaying a vigorous activity in converting AMP and GMP to the corresponding nucleoside di- and triphosphates (NDP, NTP). The underlying pattern of substrate specificity of the kinases mediating the phosphorylation of AMP and GMP and the absence of IMP kinase activity appeared to be part of a regulatory design, correlated with the role of IMP as a key intermediate in the overall purine nucleotide metabolism. This consideration prompted the present study aimed at the characterization of the different red cell NMP kinases (NTP : NMP phosphotransferases) present in the RBC. Although the data to be reported here concern specifically the NMP kinases of rabbit RBC, it should be pointed out that some observations in the literature (2,3) as well as comparative experiments done with human erythrocytes (Mager, J. and Benedikt, M., unpublished results) indicate an essential similarity in the properties of the homologous enzymic systems.

It is hoped that the information accrued from the investigation of normal pathways of nucleotide biosynthesis may contribute to a better understanding of some of the clinicopathological aspects of metabolism which form the central theme of the current colloquium.

MATERIALS AND METHODS

The various purine nucleotides were acquired from Sigma Chemical Co. (St. Louis, Mo., USA). Crystalline preparations of pyruvate kinase and lactate dehydrogenase were obtained from Boehringer Corp. (Mannheim, Germany).

The polyacrylamide gels (Bio-Gel-P) used for gel filtration as well as the hydroxylapatite preparation (Bio-Gel-HTP) were purchased from Bio-Rad Laboratories (Richmond, Calif., USA) and DEAE-Sephadex-A-50 (capacity, 3.5 mEq/g) was a product of Pharmacia (Uppsala, Sweden).

The activity of the various nucleoside monophosphokinases was measured by coupling the reaction to the conventional assay system consisting of pyruvate kinase, phosphoenol pyruvate and lactate dehydrogenase, and determining the rate of disappearance of reduced nicotinamide adenine dinucleotide (NADH) (4).

The reaction mixture contained in a final volume of 1 ml, 50 mM Tris-chloride buffer (pH 7.4), 10 mM $MgCl_2$, 80 mM KCl, 0.6 mM EDTA, 0.8 mM phosphoenolpyruvate, 0.15 mM NADH, 1 μg lactate dehydrogenase, 4 μg pyruvate kinase, 1 mM amounts of the substrates tested and a suitable amount of enzyme preparation. The reaction was initiated by the addition of either enzyme or NMP, and the decrease in optical density at 340 mμ was followed at 30 C in the recording Unicam model S-P-800B spectrophotometer. The values were corrected for the activity of the nucleoside triphosphatases by subtracting the blank values obtained in parallel runs with NMP omitted. One unit of activity was defined as the amount of enzyme catalyzing the phosphorylation of 1 μmole NMP per min and equivalent to the disappearance of 2 μmole of nicotinamide adenine dinucleotide phosphate (NADP) in the assay system. Commercial preparations of GTP and ITP contained 10 to 15% of the corresponding NDP as contaminants. In testing these substrates, therefore, the reaction mixture was preincubated for a few min until all the NDP was converted to NTP by the action of pyruvate kinase, prior to starting the reaction by addition of the NMP kinase.

By this method, linear time curves and proportional responses to varying amounts of enzyme were obtained on assaying ATP : AMP and ATP : GMP phosphotransferases. In contrast, while determining GTP-AMP phosphotransferase activity, a gradual acceleration in the rate of disappearance of NADH was observed. The seemingly aberrant kinetics were obviously due to the increasing amounts of ATP produced in the course of the reaction and the resultant interference by the ATP : AMP kinase present in excess in the enzyme source. Because of this interference, the activity of GTP-AMP phosphotransferase was measured during the first 2 min only, using an amount of enzyme sufficiently low to maintain the change of absorbance at 340 μm at about 0.03 OD units per min.

Protein was determined by the method of Lowry et al. (5).

RESULTS

Nucleoside monophosphate kinases present in red cell hemolysates. A survey of the various phosphotransferase activities in hemolysates of rabbit erythrocytes is shown in Table 1. It may be seen that apart from the highly active ATP : AMP kinase (adenylate kinase), much lower, though still appreciable, activities were also found with the GTP : AMP and ATP : GMP combinations. On the other hand, in agreement with our previous results obtained by a different experimental approach (1), no phosphorylation of IMP took place when either ATP or GTP were used as the phosphate donors. All the phosphotransferase activities were found to reside practically entirely in the stroma-free supernatant fraction of the hemolysate.

Purification procedure. All operations were carried out in the cold. In the first step, 130 ml of rabbit red cell hemolysate, prepared as previously described (1), were concentrated about threefold by stirring for 1 hr with 21 g of dry Bio-Gel-P-10, followed by filtration on a Buchner funnel. A 20 ml sample of the filtrate was applied on a DEAE-Sephadex-A-50 column (2.5 × 40 cm), previously washed with 0.01 M phosphate buffer (pH 7.4), and the same buffer was used for elution. All the hemoglobin emerged in the initial portion of the effluent (void volume) and was followed by a well defined peak containing the ATP : AMP and GTP : AMP kinases (Fraction I). The ATP : GMP kinase, on the other hand, which remained strongly adsorbed to the resin bed, was then eluted as a separate fraction (Fraction

TABLE 1. *NMP phosphotransferase activities in rabbit red cell hemolysates*

Phosphotransferase system	Activity units/ml red cells
ATP : AMP	11.30 ± 1.5
ATP : GMP	1.95 ± 0.21
ATP : IMP	0
GTP : AMP	0.94 ± 0.12
GTP : GMP	0.02 ± 0.01
GTP : IMP	0

The stroma-free hemolysate was prepared from washed rabbit erythrocytes and freed from endogenous nucleotides by treatment with charcoal, as previously described (1). The enzymic assays were carried out as detailed in Materials and Methods.

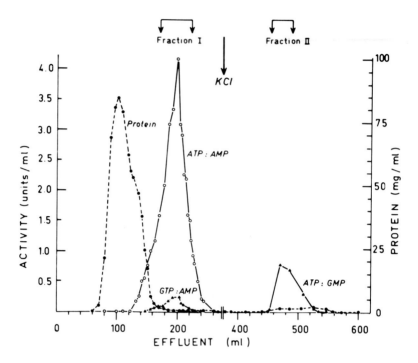

FIG. 1. Chromatographic separation of rabbit red cell NMP phosphotransferases on a DEAE-Sephadex-50-P column. The details of the procedure are given in the text. The arrow indicates the start of elution with a KCl gradient. The effluent collected in 10 ml portions was assayed for protein and NMP phosphotransferase activity (see Materials and Methods) with the following pairs of substrates: ATP : AMP, GTP : AMP and ATP : GMP. The active fractions were pooled.

II) with a gradient system consisting of 2 M KCl solution in the reservoir and 400 ml potassium phosphate buffer (pH 7.4) in the mixing chamber. A typical elution profile of the different fractions is depicted in Fig. 1.

Further purification of Fraction I was achieved by chromatography on a hydroxylapatite column (Fig. 2). A 30 ml sample of Fraction I was applied to a hydroxylapatite column (2.3 × 17 cm) followed by washing with 50 ml of 0.01M phosphate buffer (pH 7). The active fraction (Fraction Ia) was obtained by elution with a gradient system set up with 1 M phosphate buffer (pH 7) against 300 ml of 0.01M phosphate buffer (pH 7) in the mixing chamber.

Fraction II could be also further purified by use of the gel filtration technique. In a typical experiment represented by Fig. 3, 17 ml of Fraction

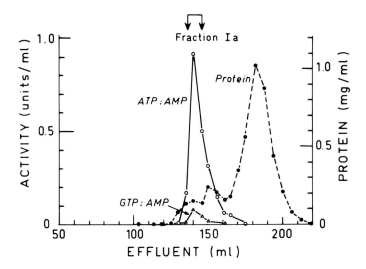

FIG. 2. Chromatography of Fraction I (AMP-specific phosphotransferases) on a hydroxylapatite column. Three ml portions were collected and assayed for protein and phosphotransferase activity with the ATP : AMP and GTP : AMP systems. The active fractions (Fraction Ia) were pooled. For other details see text.

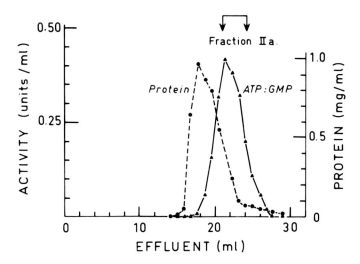

FIG. 3. Purification of Fraction II by filtration on a Bio-Gel-P-60 column. The effluent was collected in 1 ml portions, and assayed for protein and ATP : GMP phosphotransferase activity. The active fractions (Fraction IIa) were pooled. For other details see text.

TABLE 2. *Flow-sheet of the purification procedure of rabbit red cell NMP phosphotransferases*

Phosphotransferase system	Purification stage	Specific activity x 10^{-2} (units/mg protein)	Yield %	Degree of purification (times)
ATP : AMP	Bio-Gel-P-10	1.64	100	1
	Fraction I	94.40	41	58
	Fraction Ia	536.00	16	327
GTP : AMP	Bio-Gel-P-10	0.16	100	1
	Fraction I	9.34	41	58
	Fraction Ia	47.90	18	298
ATP : GMP	Bio-Gel-P-10	0.31	100	1
	Fraction II	30.00	51	99
	Fraction IIa	112.00	19	367

For experimental details see text and Fig. 1 to 3.

II were dialyzed overnight against 2 liters of distilled water with continuous magnetic stirring. The preparation was then freeze-dried, redissolved in 1 ml of 0.01M phosphate buffer (pH 7), placed on a Bio-Gel-P-60 column (0.8 × 55 cm) and the enzyme (Fraction IIa) was eluted with the same buffer.

Properties of the enzymic fractions. As evident from the summary presented in Table 2, the overall purification scheme afforded a nearly 300-fold increase in the specific activities of the various nucleoside monophosphate kinases, as well as physical separation between two enzymic fractions with strict specificities for either AMP or GMP (or the corresponding deoxyribonucleotides). The two fractions differed strikingly with regard to their phosphate donor specificities. Thus, Fraction I exhibited a broad spectrum of AMP kinase activity with a variety of different nucleoside triphosphates, whereas Fraction II showed a specific requirement for ATP in mediating the phosphorylation of GMP (Table 3). No phosphate transfer was detected with either IMP or xanthosine monophosphate (XMP), when the fractions were tested each in the presence of ATP or GTP as phosphate donors. It may also be mentioned in this connection that the AMP kinase activity of the red cell, Fraction I, conformed with respect to substrate specificity with the NTP : AMP kinase (EC 2.7.4.4.) which has been isolated from calf liver (6) except that the latter enzyme could be separated from the conventional ATP-specific AMP kinase (EC 2.7.4.3.) (7). On the other hand, enzymes similar to the ATP : GMP kinase (EC 2.7.4.8.) of Fraction II have been described in bacteria (8) and hog brain (9).

TABLE 3. *Substrate specificity of the purified NMP phosphotransferase fractions from rabbit red cell hemolysates*

Fraction Ia		Fraction IIa	
Kinase system	Relative activity %	Kinase system	Relative activity %
ATP : AMP	100	ATP : GMP	100
ATP : dAMP	16.4	ATP : dGMP	81
ATP : GMP	0	ATP : AMP	0
ATP : IMP	0	ATP : IMP	0
ATP : XMP	0	ATP : XMP	0
ATP : UMP	0	ATP : UMP	0
ATP : CMP	0	ATP : CMP	0
GTP : AMP	8.6	GTP : GMP	0
ITP : AMP	8.2	ITP : GMP	0
UTP : AMP	7.2	UTP : GMP	0
CTP : AMP	9.0	CTP : GMP	0
GTP : GMP	0	GTP : IMP	0
GTP : IMP	0	GTP : AMP	0
GTP : XMP	0	ITP : IMP	0

Experimental details as described in the text.

Although the purification procedure described, as well as some additional attempts, including adsorption of alumina Cγ gel or on carboxymethylcellulose, failed to bring about further resolution of Fraction II, the existence of distinct molecular entities was indicated by the difference in the Michaelis Menten constant (K_m) values obtained with the various systems. As shown in Table 4, the K_m for AMP, when determined with GTP as cosubstrate, was found to be about 20-fold higher than the corresponding value obtained with the ATP : AMP system. Furthermore, the GTP : AMP phosphotransferase activity was completely inhibited by sulfate ions (0.04 M $(NH_4)_2SO_4$ or 0.04 M K_2SO_4), while under the same conditions the ATP : AMP system showed only a 10 to 15% decrease in the reaction rate. Both activities were completely suppressed by 10^{-5} M *p*-hydroxymercuribenzoate.

The red cell ATP : AMP kinase did not share the property of remarkable heat stability characteristic for the congenerous muscle enzyme (myokinase) (10). The red cell enzyme was found to undergo rapid inactivation on heating at 50 C, with a half-time estimated at about 15 min. Futhermore, no significant difference in heat stability was observed when the ATP : AMP and GTP : AMP systems were compared. The pH optima of the various NMP kinase activities were within the range of 7 to 7.4.

In a search for a possible regulatory mechanism, the individual NMP phosphotransferase systems were examined with a variety of purine nucleoside mono- and triphosphates added at a 0.5 mM level. However, no appreciable alteration resulted in the reaction rates as determined by the standard procedure. Furthermore, addition of GTP or ITP at 1 mM concentration did not modify the K_m values for ATP of the ATP : AMP kinase. The latter observation was taken as additional indirect evidence in favor of the dual molecular nature of the two types of AMP kinases with distinctive specificities for the NTP cosubstrate.

DISCUSSION

The intrinsic control mechanisms governing the biosynthesis of purine nucleotides appear to be designed to obviate excessive accumulation of AMP and GMP in the cell and to promote the formation of the corresponding NTP. AMP and GMP are known to exert a strong feedback inhibition on the specific pyrophosphorylases mediating their synthesis from preformed purine bases and 5-phosphoribosyl-1-pyrophosphate (PRPP), (11, 12). Thus, the phosphorylation of AMP and GMP by the NMP kinases may be viewed in this context as a regulatory device enabling the cell to escape the retroinhibitory pressure of the NMP and to ensure the continued operation of the biosynthetic pathways, which otherwise would tend to come to an early halt. Furthermore, the conversion of the NMP to the corresponding diphosphate derivatives enhances their "metabolic stability" by rendering them unavailable to the splitting action of the 5'-nucleotidase. On the other hand, the accumulating NDP offer an additional means for restraining nucleotide synthesis by inducing a potent inhibition of the synthetase catalyzing the formation of PRPP (13) and thereby curtailing the supply of this essential precursor. The latter hurdle may be overcome by further phosphorylation of ADP and GDP to ATP and GTP, mediated in the RBC by the energy-yielding glycolytic reactions and the highly active NDP kinase (EC 2.7.4.6.), respectively (14).

Consonant with the above considerations appears to be the negative outcome of the search for a feedback control of the NMP kinases, suggesting that the activity of these enzymes is governed solely by their inherent kinetic properties and the availability of their substrates. It also seems reasonable to surmise in the light of this conclusion that the characteristic preponderance of the normal intracellular pool of adenine nucleotides, with a nearly tenfold excess over the guanine nucleotide level, may be at least

TABLE 4. K_m values of the red cell NMP phosphotransferase

Kinase system	Substrate	K_m (M)
ATP : AMP	ATP	7×10^{-5}
	AMP	5.5×10^{-5}
GTP : AMP	GTP	3×10^{-4}
	AMP	1×10^{-3}
ATP : GMP	ATP	8×10^{-4}
	GMP	4×10^{-5}

Purified enzyme preparations (Fraction Ia and IIa) were used.

partly determined by the similar numerical ratio of the affinity constants for ATP of the respective kinases (Table 4).

The peculiar substrate specificity of the NMP kinases and, notably, the absence of a kinase capable of catalyzing phosphorylation of IMP and XMP seem to be consistent with the crucial role of these NMP in the intermediary metabolism of purine nucleotides (in both synthesis and catabolism), since it is evident that their conversion to higher phosphorylated derivatives would be incompatible with their intermediary functions. Consequently, IMP and XMP, unless converted to AMP or GMP, are destined to be constantly exposed to the hydrolytic attack of 5'-nucleotidase and, therefore, liable to rapid degradation.

The enhanced catabolic potential of IMP is reflected *in vivo* in its extremely short life span observed in the circulating red cell and contrasting with the much slower renewal rates of the adenine and guanine nucleotides (15). This regulatory arrangement seems to be instrumental in channeling the overall catabolism of the red cell purine nucleotides towards hypoxanthine (and to a minor extent towards xanthine) which may then either reenter a new cycle of nucleotide synthesis or diffuse out of the cell and continue in the liver its path of catabolism to uric acid and allantoin.

REFERENCES

1. HERSHKO, A., RAZIN, A., SHOSHANI, T. and MAGER, J. Turnover of purine nucleotides in rabbit erythrocytes. II. Studies *in vitro*. Biochim. biophys. Acta (Amst.) **149**: 59, 1967.
2. KASHKET, S. and DENSTEDT, O. F. The metabolism of the erythrocyte. XV. Adeny-. late kinase of the erythrocyte. Canad. J. Biochem. **36**: 1057, 1958.
3. CARLETTI, P. and BUCCI, E. Adenylate kinase of mammalian erythrocytes. Biochim biophys. Acta (Amst.) **38**: 45, 1960.

4. ADAM, H.   Nulceotidspezifität glycolytischer Kinasen; Enzymatische Bestimmung des Adenylsäueresystems. *Biochem. Z.* **335**: 25, 1961.
5. LOWRY, O. H., ROSEBROUGH, N. J., FARR, A. L. and RANDALL, R. G.   Protein measurement with the Folin phenol reagent. *J. biol. Chem.* **193**: 265, 1951.
6. HEPPEL, L. A., STROMINGER, J. L. and MAXWELL, E. S.   Nucleoside monophosphate kinases. II. Transphosphorylation between adenosine monophosphate and nucleoside triphosphate. *Biochim. biophys. Acta (Amst.)* **32**: 422, 1959.
7. CHIGA, M., ROGERS, A. E. and PLAUT, G. E. W.   Nucleotide transphosphorylases from liver. II. Purification and properties of a 6-oxypurine nucleoside triphosphate-adenosine monophosphate transphorylase from swine liver. *J. biol. Chem.* **236**: 1800, 1961.
8. OESCHGER, M. P. and BESSMAM, M. J.   Purification and properties of guanylate kinase from Escherichia coli. *J. biol. Chem.* **241**: 5452, 1966.
9. MEICH, R. P. and PARKS, R. E., Jr.   Adenosine triphosphate: guanosine monophosphate phosphotransferase. Partial purification and substrate specificity. *J. biol. Chem.* **240**: 351, 1965.
10. COLOWICK, S. P. and KALCKAR, H. M.   The role of myokinase in transphosphorylations. I. The enzymatic phosphorylations of hexoses by adenylpyrophosphate. *J. biol. Chem.* **148**: 117, 1943.
11. RAZIN, A.   Interrelations in the metabolism of purine nucleotides of rat blood cells. Ph. D. Thesis, Jerusalem, 1967.
12. HENDERSON, J. F.   Kinetic properties of hypoxanthine-guanine and adenine phosphoribosyltransferases. *Fed. Proc.* **27**: 1053, 1968.
13. HERSHKO, A., RAZIN, A. and MAGER, J.   Regulation of the synthesis of 5-phosphoribosyl-1-pyrophosphate in intact red blood cells and in cell-free preparations. *Biochim. biophys. Acta (Amst.)* **184**: 64, 1969.
14. MOURAD, N. and PARKS, R. E.   Erythrocytic nucleoside diphosphokinase. II. Isolation and kinetics. *J. biol. Chem.* **241**: 271, 1966.
15. MAGER, J., HERSHKO, A., ZEITLIN-BECK, R., SHOSHANI, T. and RAZIN, A.   Turnover of purine nucleotides in rabbit erythrocytes. I. Studies *in vivo*. *Biochim. biophys. Acta (Amst.)* **149**: 50, 1967.

EVIDENCE FOR AN ABNORMAL RED CELL POPULATION IN REFRACTORY SIDEROBLASTIC ANEMIA*

ISAAC BEN-BASSAT, FRIDA BROK-SIMONI
and BRACHA RAMOT

Department of Hematology, Government Hospital, Tel-Hashomer and Tel Aviv
University Medical School, Tel Aviv, Israel

There has been considerable interest in the many aspects of the ineffective erythropoiesis which characterizes chronic sideroblastic refractory anemia (1-6). Although the basic defect is still unknown there is evidence from different studies that several red cell populations exist in this disease (2-7).

As part of a larger study concerning the *in vivo* aging pattern of various red cell enzymes in different hematologic diseases, we have studied a group of patients with chronic refractory anemia and observed in all of them the appearance of an abnormal red cell population which, so far, seems typical of this disorder. The present report describes these findings.

SUBJECTS AND METHODS

The 11 patients studied had typical clinical and laboratory features of chronic refractory sideroblastic anemia. In eight, no other disease was discovered, while in three patients another disease (chronic myeloid leukemia in two and miliary tuberculosis in one) was found to be associated with the hematologic disorder.

The enzymatic studies were performed as a rule several months after transfusions. In three patients, however, only a few weeks had lapsed. Almost all were examined at least twice and a few were studied on several occasions. A constant and reproducible pattern could be seen at each examination.

Blood from normal laboratory personnel and nonhematological patients served as control. Fourteen patients with hemolytic diseases of various

---

* Part of this study has been published in *Blood* **35**: 453, 1970.

etiologies were also studied to determine the effect of reticulocytosis on the enzymatic activities. In addition to routine hematological and laboratory investigations, the following methods were used: osmotic fragility curve of fresh whole blood was recorded with a fragiligraph (8) and density distribution curves were determined according to Danon and Marikovsky (9). Red cells were separated into age groups by the differential flotation method, using phthalate ester mixtures of predetermined specific gravity, as published previously (10). The white cells were separated from the red cells by passing the blood through a phthalate mixture of sp gr 1.062. In each blood sample we separated the lightest 1% cells (first light fraction) and the next 2% fraction of light cells (second light fraction) as well as the 1% densest cells (first dense fraction) and the next 2% cells (second dense fraction). In some cases the third sequential fraction of light cells amounting approximately to 5% (third light fraction) was also separated. Reticulocyte counts were performed on each fraction. The activities of glucose-6-phosphate dehydrogenase (G6PD) (11), 6-phosphogluconic dehydrogenase (6PGD) (12), hexokinase (HK) (13) and pyruvate kinase (PK) (14) were determined on each fraction and on whole blood, according to the cited methods. The ATP content was determined by the phosphoglycerate kinase reaction (15) and erythrocyte glutamic oxaloacetic transaminase (GOT) activity was measured by the colorimetric method (16) modified for hemolysates.

RESULTS

The results of the autohemolysis test, osmotic fragility of the fresh red cells and the density distribution curves were generally within normal limits and did not provide evidence for the existence of an abnormal population of red cells.

As can be seen in Table 1, the mean activities of red cell G6PD, 6PGD and HK in whole blood were high in the refractory anemia group when compared to normal subjects. ATP levels, however, were identical. When the enzymatic activities and ATP content were examined in the two densest fractions, no significant differences were found. On the other hand, when the 1% lightest cells were examined it was found that in refractory anemia these cells had a very high enzymatic activity, 2.5 to 5.5 times the levels found in whole blood, compared to 1.6 to 2.5 in normal blood. The lightest cells in refractory anemia were especially rich in HK activity which reached 550% compared to 200% in normal cells. The differences in the second light fractions were much less striking but still evident. However, the third light

TABLE 1. *Enzymatic activities, ATP content and reticulocyte counts of the various fractions in refractory anemia and normal subjects*

			Lightest				Whole blood		Densest			
			1st fraction		2nd fraction				2nd fraction		1st fraction	
		Determinations	Mean	SD	Mean	SD	Mean	SD	Mean	SD	Mean	SD
G6PD [a]	RA	20	10.6	±6.2	5.0	±2.1	3.6	±0.4	2.2	±0.9	1.9	±0.9
	N	40	4.4	±0.3	3.4	±0.9	2.6	±0.5	2.1	±1.0	1.3	±0.3
			$P < 0.001$		$P < 0.01$		$P < 0.001$		NS		$P < 0.01$	
6PGD [a]	RA	5	12.7	±4.8	7.0	±2.7	4.8	±1.2	3.8	±0.7	2.9	±0.8
	N	40	4.6	±1.1	—	—	2.8	±0.6	2.2	±0.6	2.1	±0.6
			$P = 0.02$				$P = 0.02$		$P < 0.01$		NS	
HK [a]	RA	20	2.50	±0.41	0.72	±0.41	0.45	±0.20	0.25	±0.10	0.13	±0.10
	N	40	0.57	±0.02	0.32	±0.12	0.28	±0.04	0.21	±0.07	0.14	±0.03
			$P < 0.001$		$P < 0.001$		$P < 0.001$		NS		NS	
PK [a]	RA	7	16.1	±6.0	8.2	±2.1	6.0	±1.8	3.8	±0.9	2.9	±0.6
	N	15	8.8	±1.5	—	—	5.1	±1.0	3.3	±1.1	2.3	±1.1
			$P = 0.02$				NS		NS		NS	
ATP [b]	RA	6	2.80	±0.83	1.70	±0.58	1.10	±0.27	0.64	±0.20	0.47	±0.40
	N	12	2.45	±0.70	1.68	±0.33	1.10	±0.28	0.55	±0.24	0.34	±0.18
			NS		NS		NS		NS		NS	
Retic [c]	RA	20	10.3	±3.8	5.0	±2.8	3.1	±1.0	0.6	±0.5	0.4	±0.1
	N	10	4.1	±0.3	—	—	0.6	±0.1	0.1	±0.2	0	± 0
			$P < 0.001$				$P < 0.001$		$P < 0.01$			

[a] = μmole/gHb per min
[b] = μmole/ml Hb
[c] = %

RA = Refractory anemia
N = Normal
NS = Not significant

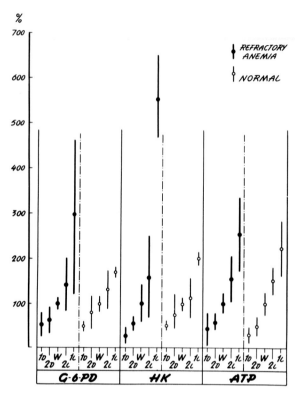

FIG. 1. G6PD and HK activities and ATP content of the various red cell fractions in refractory anemia and normals. The results are expressed as percent of the activity found in whole blood.

1D = first dense fraction; 2D = second dense fraction; W = whole blood; 2L = second light fraction; 1L = first light fraction.

fractions, whenever isolated, had enzymatic activities very similar to whole blood. In contrast to the enyzmatic findings, the ATP levels in all the refractory anemia fractions were parallel with, if not identical to, those in the control subjects. These experimental results are expressed graphically in Fig. 1 where G6PD and HK activities and ATP contents are given as percentages of whole blood.

In order to exclude the possibility that the high enzymatic activities in the refractory anemia lightest cells are due to their higher reticulocyte contents (10.3% vs. 4.1% in normal 1% lightest cells), the red cells of 14 cases of hemolytic anemias of various etiologies were separated into age groups. It

TABLE 2. *Enzymatic activities and reticulocyte counts of the various fractions in refractory and hemolytic anemias*

		Determinations	Lightest 1st fraction Mean	SD	Lightest 2nd fraction Mean	SD	Whole blood Mean	SD
G6PD[a]	RA	20	10.6	±6.2	5.0	±2.1	3.6	±0.4
	HA	14	5.6	±1.8	4.1	±1.4	2.8	±1.4
			$P < 0.01$		NS		NS	
HK[a]	RA	20	2.50	±0.41	0.72	±0.41	0.45	±0.20
	HA	14	1.10	±0.50	0.81	±0.45	0.43	±0.15
			$P < 0.001$		NS		NS	
Retic[b]	RA	20	10.3	±3.8	5.0	±2.8	3.1	±1.0
	HA	14	39.8	±9.0	20.4	±4.0	12.5	±3.0
			$P < 0.001$		$P < 0.001$		$P < 0.001$	

RA = Refractory anemia   [a] = μmole/gHb per min
HA = Hemolytic anemia   [b] = %
NS = Not significant

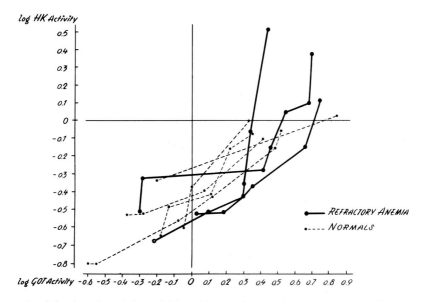

FIG. 2. A log-log plot of the activities of HK and GOT, determined as μmole/gHb per min, in the various density fractions in refractory anemia and normal controls. Each point represents a density fraction: the extreme points are the 1% lightest (right) and 1% densest (left) fractions.

can be clearly seen from Table 2 that even a much higher reticulocytosis does not cause such elevations of the enzymatic activities.

As GOT is considered the most reliable age-dependent red cell enzyme (17), its activity was determined in the various red cell fractions of three patients and six control subjects. Fig. 2 is a log-log plot of the GOT activity vs. the HK activity of each red cell fraction. It demonstrates again the presence in refractory anemia of a population with very high HK activity, out of proportion to the number of reticulocytes in this cell fraction. The precipitous drop of HK activity in the following density fraction is also seen.

DISCUSSION

The 11 patients studied had the typical features of chronic sideroblastic refractory anemia as defined by other authors (1–6). The basic defect in this disease is still unknown. Impaired hemoglobin synthesis, defective red cell maturation, multiple enzymatic deficiencies and a malignant neoplastic process have been suggested as etiologic factors. Waltuch et al. (18) reported a defect in heme synthesis, a finding which has been reported previously in similar patients in the DiGuglielmo syndrome (19, 20). Lewis et al. (21) and Cooper et al. (22) found increased agglutinability of the red cells in sideroblastic anemia by anti-i and anti-I cold antibodies. Dreyfus et al. (7) found significant quantitative changes in one or more red cell antigens. Moreover, there were several red cell populations differing in their concentration of red cell antigens.

The different natural history in many cases, especially in relation to the development of leukemia, the multiplicity of findings reported, and their similarity in the primary and secondary forms indicate that we are dealing with a heterogeneous group that has certain features in common. Our findings seem to be only one facet of the complex metabolic defect found in this disease. The method employed in this study allowed a reliable separation of the red cells into age groups as evidenced by the gradients of the reticulocyte counts, GOT and other enzymatic activities. Thus we can assume that the abnormal lightest fraction isolated in refractory anemia, is a population of very young cells shortly after their release from the bone marrow. These cells have higher enzymatic activities than reticulocytes and do not have their staining properties. At the same time their ATP content is low, compared with the enzymatic activities, and similar to that of normal young cells. Some of these cells are probably sequestered shortly after being released, due to their low surface charge, as are normal senescent cells (23).

So far we have not found such a red cell population in any other hematologic disorder, including megaloblastic anemias and thalassemia — conditions in which ineffective hematopoiesis also occurs. We have no explanation at present for our findings. Possibly they are reflections of a more basic underlying defect in proliferation and maturation of the red cell precursors which leads to the release of abnormal cells, possibly via skipped generations, due to "marrow stress."

We wish to thank Mrs. F. Holtzman for her technical assistance. The statistical evaluation was generously done by Miss P. Weisskopf, M.Sc.

REFERENCES

1. BJÖRKMAN, S. E. Chronic refractory anemia with sideroblastic bone marrow. A study of four cases. *Blood* **11**: 250, 1956.
2. DACIE, J. V., SMITH, M. D., WHITE, J. C. and MOLLIN, D. L. Refractory normoblastic anemia: a clinical and haematological study of seven cases. *Brit. J. Haemat.* **5**: 56, 1969.
3. VILTER, R. W., JARROLD, T., WILL, J. J., MUELLER, J. F., FRIEDMAN, B. I. and HAWKINS, V. R. Refractory anemia with hyperplastic bone marrow. *Blood* **15**: 1, 1960.
4. MACGIBBON, B. H. and MOLLIN, D. L. Sideroblastic anemia in man: observations on seventy cases. *Brit. J. Haemat.* **11**: 59, 1965.
5. DAMESHEK, W. Sideroblastic anemia: Is this a malignancy? *Brit. J. Haemat.* **11**: 52, 1965.
6. BARRY, W. E. and DAY, H. J. Refractory sideroblastic anemia. Clinical and hematologic study of ten cases. *Ann. intern. Med.* **61**: 1029, 1964.
7. DREYFUS, B., SULTAN, C., ROCHANT, H., SALMON, CH., MANNONI, P., CARTRON, J. P., BOIVIN, P. and GALAND, C. Anomalies of blood group antigens and erythrocyte enzymes in two types of chronic refractory anemia. *Brit. J. Haemat.* **16**: 303, 1969.
8. DANON, D. A rapid micro method for recording red cell osmotic fragility by continuous decrease of salt concentration. *J. clin. Path.* **16**: 377, 1963.
9. DANON, D. and MARIKOVSKY, Y. Determination of density distribution of red cell population. *J. Lab. clin. Med.* **65**: 668, 1964.
10. BROK, F., RAMOT, B., ZWANG, E. and DANON, D. Enzyme activites in human red blood cells of different age groups. *Israel J. med. Sci.* **2**: 291, 1966.
11. KORNBERG, A. and HORECKER, B. L. Glucose-6-phosphate dehydrogenase, in: Colowick, S. P. and Kaplan, N. O. (Eds.) "Methods in Enzymology." New York, Academic Press, 1955, v. I, p. 323.
12. HORECKER, B. L. and SMYRNIOTIS, P. Z. 6-phosphogluconic dehydrogenase, in: Colowick, S.P. and Kaplan, N.O. (Eds.), "Methods in Enzymology." New York, Academic Press, 1955, v. I, p. 323.
13. GRIGNANI, F. and LÖHR, G. W. Über die Hexokinase in menschlichen Blutzellen. *Klin. Wschr.* **38**: 796, 1960.
14. BOCK, H. E., WALLER, H. D., LÖHR, G. W. and KARGES, O. Besonderheiten im Fermentgehalt von Megalocyten. *Klin. Wschr.* **36**: 151, 1958.
15. HOLLOWAY, B. W. The determination of adenosine triphosphate by means of phosphoglyceric acid kinase. *Arch. Biochem.* **52**: 33, 1954.
16. REITMAN, S. and FRANKEL, S. A colorimetric method for the determination of serum glutamic oxaloacetic and glutamic pyruvic transaminases. *Amer. J. clin. Path.* **28**: 56, 1955.
17. BARTOS, H. R. and DESFORGES, J. F. Enzymes as erythrocyte age reference standards *Amer. J. med. Sci.* **254**: 862, 1967.

18. WALTUCH, G., LANZEROTTI, A. K. and SCHRIER, S. L.   Marrow defect in idiopathic ineffective erythropoiesis. *Ann. intern. Med.* **68**: 1005, 1968.
19. STEINER, M., BALDINI, M. and DAMESHEK, W.   Heme synthesis in "refractory" anemia with ineffective erythropoiesis. *Blood* **22**: 810, 1963.
20. NECHELES, T. F. and DAMESHEK, W.   The DiGuglielmo syndrome: studies in hemoglobin synthesis. *Blood* **29**: 550, 1967.
21. LEWIS, S. N., DACIE, J. V. and TILLS, D.   Comparison of the sensitivity to agglutination and haemolysis by a high-titre cold antibody of the erythrocytes of normal subjects and of patients with a variety of blood diseases including paroxysmal nocturnal haemoglobinuria. *Brit. J. Haemat.* **7**: 64, 1961.
22. COOPER, A. G., HOFFBRAND, A. V. and WORLLEDGE, S. M.   Increased agglutinability by anti i of red cells in sideroblastic and megaloblastic anaemia. *Brit. J Haemat.* **15**: 381, 1968.
23. DANON, D.   Biophysical aspects of red blood cells ageing. *11th Cong. Int. Soc. Haematol. Australia*, 1966, p. 394.

DISCUSSION

B. Ramot (*Israel*): We have studied the glycolytic and pentose pathway enzymes in iron deficiency anemia and thalassemia and have obtained results similar to those of Dr. Mazza. The very high enzyme levels observed after iron therapy are the result of the appearance of a young cell population, partly as a result of stress erythropoiesis. We were able to prove this using a density flotation method with phthalate esters and enzyme determinations in each density fraction.

E. Goldschmidt (*Israel*): I have a remark as well as a question. You mentioned, if I understood you correctly, that you separate the upper fraction of RBC by the phthalate method of Danon and Marikovsky (1964), and obtain less than 1% of the youngest cells. We have made some preliminary experiments with this method and have never succeeded in separating less than the upper 10%. We have tried various methods of getting the upper RBC off the phthalate mixture: In a large test tube or in the 0.6 ml test tubes which go with the Spinco Microfuge (ca. 12,000 rpm) the upper fraction may be pipetted off with some skill. As a rule this results either in considerable loss of RBC or in contamination with a great deal of phthalate. It is more useful to cut the upper fraction off clean. This may be done by means of scissors with the microhematocrit capillary tubes originally recommended by Danon and Marikovsky, or by means of a sharp razor blade with the plastic test tube supplied with the Spinco Microfuge. By neither method have we succeeded in obtaining young fractions as small as yours. My question concerns the calculations that you applied to your rough enzyme data. When calculating enzyme activities of all types of RBC, but especially of pathological populations which may vary widely in hemoglobin content and in cell size and shape, it is advisable to calculate enzyme activity by three different methods: 1) per hemoglobin (the most commonly applied calculation), 2) per cell and 3) per protein.

B. Ramot: We have calculated enzyme activity per gram Hb because it is difficult to obtain accurate counts without a coulter counter (CC). Even if you would assume a mean corpuscular Hb concentration of 50% of normal, you would still have enzyme levels in iron deficiency that are much too high. As to the cell separation, my only answer is Tel Hashomer is not far away and you are invited to see the procedure yourself.

E. Goldschmidt: We do not use a CC either, but count RBC by the classical method in a counting chamber under the microscope. I have worked in a laboratory concerned with problems similar to the ones you are investigating and they used a CC for both the white and the red cell count. The attention to this delicate instrument was quite time consuming even in the hands of a first rate senior scientist. I have been told that it is not advisable to use the CC for a cell population liable to exhibit wide variance in size since the instrument is sensitive to both cell size and cell number. But I have no personal experience in the use of this instrument.

B. Ramot: Cells separated after phthalate are hard to count. The error is so great that I do not rely on such results.

SESSION V

Chairmen: R. Kleihauer, *West Germany*
J. M. Schwartz, *USA*

Participants: N. Bashan, *Israel*
I. Ben-Bassat, *Israel*
K. G. Blume, *West Germany*
D. Busch, *West Germany*
R. Chayoth, *Israel*
R. W. Hoffbauer, *West Germany*
G. W. Loehr, *West Germany*
S. W. Moses, *Israel*
B. Ramot, *Israel*
M. Shchory, *Israel*

KINETIC PROPERTIES OF PYRUVATE KINASE AND PROBLEMS OF THERAPY IN DIFFERENT TYPES OF PYRUVATE KINASE DEFICIENCY

D. BUSCH, R. W. HOFFBAUER, K. G. BLUME and
G. W. LOEHR

Medical Clinic, University of Freiburg im Breisgau, West Germany

Pyruvate kinase (PK) deficiency is the most common defect in the glycolytic pathway of erythrocytes occurring in hereditary nonspherocytic hemolytic anemia, and more than 120 cases have so far been published. However, this deficiency does not represent a homogeneous group. Waller and Löhr (1), Paglia et al. (2), Boivin et al. (3), Sachs et al. (4) and Fusco et al. (5) reported on PK with an altered affinity to the substrate phosphoenolpyruvate (PEP). The total enzyme activity in the latter cases was either normal or increased (5). The majority of cases, however, have diminished enzyme activity (6, 7). The degree of reduction observed varied in magnitude and showed a bimodal distribution (Fig. 1) (7). The first group, formerly called type A, has an enzyme activity of about 5 to 30% of normal, whereas the second group, type B, shows 50 to 70% of normal enzyme activity. Family studies indicate an interfamilial, but not an intrafamilial variability.

In this communication, differences in respect to fructose-1,6-diphosphate (FDP) activation kinetics of the enzyme are presented as an essential discriminating feature in types A and B PK deficiency. Further kinetic, metabolic and electrophoretic studies are described.

Electrophoresis of PK from human erythrocytes. Leukocyte-free preparations of erythrocytes were obtained by cotton wool filtration (8). PK from normal human erythrocytes was separated on cellulose acetate by high voltage electrophoresis in a phosphate buffer, pH 5.3, into three bands: A anodic, B and C cathodic (Fig. 2) (9). The staining procedure was a coupled PK-lactic dehydrogenase (LDH) reaction according to Fellenberg et al. (10). Under these conditions the C band was not detectable in either type of PK deficiency. The same C band was markedly decreased in the hemoly-

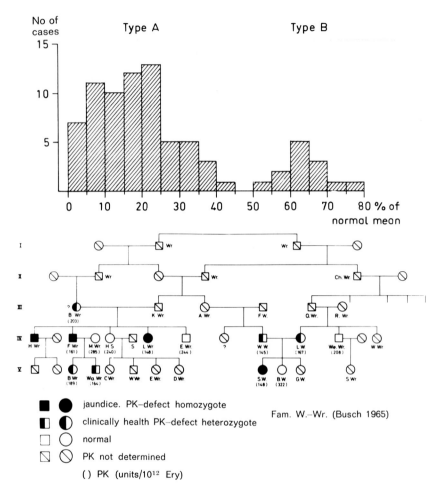

FIG. 1. Distribution of PK activity in type A- and B-deficient patients with pedigrees of families. PK activity in μmole of PEP converted per 10^{12} erythrocytes/min at 25 C, pH 7.5 (7).
Normal mean value = 265 units (SD = 35 units, normal range 195 to 335 units).

sate of the parents of a PK-deficient patient of type B (S.W.). Surprisingly, this fraction C, which was not seen on electrophoresis, could be demonstrated by ammonium sulfate precipitation as will be shown later. This is explained by a different inactivation rate of the fraction by enzyme aging.

Kinetic properties of PK from human erythrocytes. The activity of PK is dependent on the concentration of the enzyme itself, substrates such as

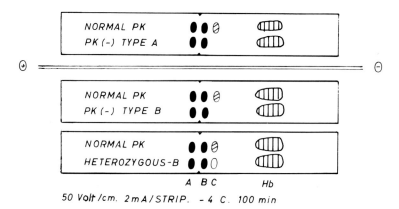

FIG. 2. High voltage electrophoresis (cellulose acetate) of erythrocyte PK from normal individuals and from patients with type A and B PK deficiency. Electrophoresis buffer system: 0.01M KH_2PO_4/Na_2HPO_4, pH 5.3, EDTA, 4×10^{-4} M, 2-mercaptoethanol 2×10^{-6} M.
Staining solution: triethanolamine buffer 6.0×10^{-2} M, pH 7.5, KCl 8.0×10^{-2} M, $MgCl_2$ 6.5×10^{-3} M, ADP 7.0×10^{-4} M, PEP 2.3×10^{-3} M, NADH 4.0×10^{-4} M, LDH 0.45 units/ml.

PEP, ADP, pyruvate, ATP as well as FDP, K^+, Mg^{++} and $(NH_4)^+$-ions and inhibitors as ATP in high concentrations and copper. Hess et al. (11,12) were the first to detect allosteric properties of yeast PK. The activation of the erythrocyte PK by FDP was first reported by Koler et al. (13). Allosteric activation was observed with the ligands PEP, ADP, FDP and potassium ions.

Hill-plots given in Fig. 3, illustrate the dependence of PK-reaction velocity on the concentrations of PEP, ADP and FDP. In the absence of FDP, a sigmoid saturation curve for PEP is obtained with half maximal velocity ($K_{\frac{1}{2}}$) at 3.4×10^{-4} M and with a Hill exponent n_H 1.8. FDP transforms the sigmoid curve into a hyperbolic form, lowering the values for $K_{\frac{1}{2}}$ and n_H. When the PEP level is below PK saturation, the ADP-dependent velocity curve shows sigmoidicity too. It is transformed to a hyperbolic one at saturating PEP concentrations, giving $K_{\frac{1}{2}}$ of 4.0×10^{-4} M and n_H 1.0. The FDP activation curve of the enzyme in the absence of PEP saturation is also sigmoid with a very low $K_{\frac{1}{2}}$ of 0.6×10^{-7} and n_H 1.7.

PK fractions from hemolysates of normal erythrocytes. The method of Aebi et al. (14) was used for purification of PK. A relative purification of the three electrophoretic fractions of the hemolysate was obtained by

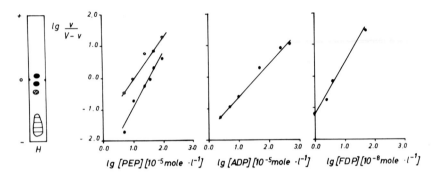

FIG. 3. Kinetic properties of PK in normal hemolysate after Sephadex G-25 filtration with 1.5×10^{-1} M KCl to eliminate glycolytic intermediates (column 90×8 mm, $V_0 = 2.0$ ml). Electrophoresis and dependence of reaction velocity on concentrations of PEP, ADP, FDP; Hill plots; $K_{\frac{1}{2}}$: molar concentrations of a ligand at half maximal activity.

Hill exponent $\quad n_H = \dfrac{\Delta \lg \dfrac{v}{V-v}}{\Delta \lg\ [L]}$

L = ligand

Assay conditions: triethanolamine buffer 6.0×10^{-2} M, KCl 8.0×10^{-2} M, MgCl$_2$ 6.5×10^{-3} M, ADP 7.0×10^{-4} M, PEP 2.3×10^{-3} M, NADH 2.0×10^{-4} M, LDH 0.45 units/ml; start with PEP at 25 C; assays with FDP started after preincubation with FDP for 5 min at 25 C.

ammonium sulfate precipitation (saturation of 0.20 to 0.38, 0.38 to 0.45 and 0.45 to 0.70) (Fig. 4). These fractions were free of hemoglobin. The relative purification was similar to that obtained by Bigley et al. (15). In Fig. 4 to 6, Hill plots for the three electrophoretic fractions of PK (A, B and C) from normal individuals are depicted. The values of $K_{\frac{1}{2}}$ and n_H for PEP in the presence of saturating concentrations of FDP do not show significant differences in the three fractions, being 0.6 to 0.8×10^{-7} and n_H 0.8 to 0.9. In the absence of FDP there is a trend for decreasing values of $K_{\frac{1}{2}}$ and n_H in the direction A → B → C approaching the values obtained in the presence of FDP. A similar trend was observed for the values of $K_{\frac{1}{2}}$ and n_H for ADP within the same fractions. However, the amplitude of this effect is much smaller because of the saturating concentration of PEP. The Hill exponents for FDP are nearly equal for A, B and C fractions.

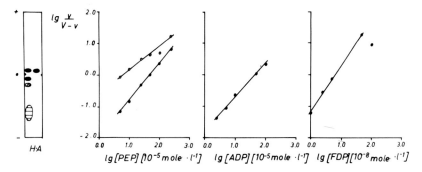

- [PEP];[FDP] ≅ 0.0·10⁻⁴ mole·l⁻¹ : K₁/₂ = 4.8·10⁻⁴ mole·l⁻¹ ; n_H = 1.2
- [PEP];[FDP] ≅ 5.0·10⁻⁴ mole·l⁻¹ : K₁/₂ = 0.6·10⁻⁴ mole·l⁻¹ ; n_H = 0.8
- [ADP];[FDP] ≅ 0.0·10⁻⁴ mole·l⁻¹ : K₁/₂ = 4.4·10⁻⁴ mole·l⁻¹ ; n_H = 1.1
- [FDP];[PEP] ≅ 5.0·10⁻⁵ mole·l⁻¹ : K₁/₂ = 0.6·10⁻⁷ mole·l⁻¹ ; n_H = 1.5

FIG. 4. Kinetic properties of PK fraction A isolated from normal individual. Sephadex G-100 filtration of hemolysate. Elution buffer: KCl 1.5×10^{-1} M, MgCl₂ 5.0×10^{-3} M, Na₂ HPO₄/KH₂PO₄ 1.0×10^{-2} M, pH 6.2, 2-mercaptoethanol 2.0×10^{-6} M; column 560 × 16 mm, V_o = 32 ml, flow rate 18 ml/cm² per hr; collection of eluate V_o --→ front of hemoglobin.
(NH₄)₂SO₄ precipitation at 0.20 to 0.38 saturation (A); electrophoresis and Hill plots (for assay conditions see Fig. 3).

The $K_{\frac{1}{2}}$ values of FDP activation are similar in spite of slightly different constants shown in Fig. 4 to 6. These differences may be the consequence of experimental conditions only, as the FDP activation factor is different in the fractions A, B and C at equal PEP concentrations used. For the same reason it is not quite obvious that FDP really activates fraction C: the experimental results available today are inconsistent in this respect. Altogether, the PEP and ADP kinetics of the three fractions of normal pyruvate kinase are different in the absence of FDP; in an activated state the kinetic properties of normal PK fractions A, B and C, may point to similarity. Preliminary studies of the partially purified fractions in the presence of allosteric activators show some degree of convertibility among the three fractions.

PK obtained from hemolysates of type B-deficient erythrocytes. The Hill plots of the reaction velocity and the concentrations of PEP, ADP and FDP obtained with the unfractionated hemolysate from patient S.W., belonging to type B PK deficiency, are presented in Fig. 7. The kinetic properties of this abnormal enzyme did not show major differences with respect to sigmoidicity of the saturation curves without allosteric activation

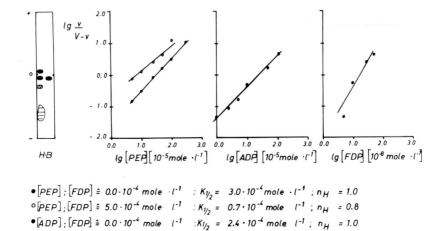

- $[PEP]$; $[FDP] \doteq 0.0 \cdot 10^{-4}$ mole $\cdot l^{-1}$: $K_{1/2} = 3.0 \cdot 10^{-4}$ mole $\cdot l^{-1}$; $n_H = 1.0$
- ○ $[PEP]$; $[FDP] \doteq 5.0 \cdot 10^{-4}$ mole l^{-1} : $K_{1/2} = 0.7 \cdot 10^{-4}$ mole l^{-1} ; $n_H = 0.8$
- $[ADP]$; $[FDP] \doteq 0.0 \cdot 10^{-4}$ mole l^{-1} : $K_{1/2} = 2.4 \cdot 10^{-4}$ mole l^{-1} ; $n_H = 1.0$
- $[FDP]_i[PEP] \doteq 5.0 \cdot 10^{-5}$ mole l^{-1} : $K_{1/2} = 1.7 \cdot 10^{-7}$ mole l^{-1} ; $n_H = 1.6$

FIG. 5. Kinetic properties of PK fraction B isolated from a normal individual. Sephadex G-100 filtration of hemolysate (for conditions see Fig. 4).
$(NH_4)_2SO_4$ precipitation at 0.38 to 0.45 saturation (B); electrophoresis and Hill plots (for assay conditions, see Fig. 3).

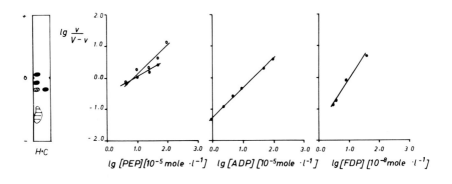

- $[PEP]$; $[FDP] \doteq 0.0 \cdot 10^{-4}$ mole l^{-1} : $K_{1/2} = 1.0 \cdot 10^{-4}$ mole $\cdot l^{-1}$; $n_H = 0.6$
- ○ $[PEP]$; $[FDP] \doteq 5.0 \cdot 10^{-4}$ mole l^{-1} : $K_{1/2} = 0.8 \cdot 10^{-4}$ mole $\cdot l^{-1}$; $n_H = 0.9$
- $[ADP]$; $[FDP] \doteq 0.0 \cdot 10^{-4}$ mole l^{-1} : $K_{1/2} = 2.3 \cdot 10^{-4}$ mole $\cdot l^{-1}$; $n_H = 0.9$
- $[FDP]$; $[PEP] \doteq 5.0 \cdot 10^{-5}$ mole l^{-1} : $K_{1/2} = 1.1 \cdot 10^{-7}$ mole $\cdot l^{-1}$; $n_H = 1.5$?

FIG. 6. Kinetic properties of PK fraction C isolated from a normal individual. Sephadex G-100 filtration of hemolysate (for conditions, see Fig. 4).
$(NH_4)_2SO_4$ precipitation at 0.45 to 0.70 saturation (C); electrophoresis and Hill plots (for assay conditions, see Fig. 3).

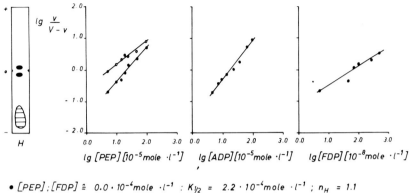

- • [PEP];[FDP] ≅ 0.0 · 10⁻⁴ mole ·l⁻¹ : $K_{1/2}$ = 2.2 · 10⁻⁴ mole ·l⁻¹ ; n_H = 1.1
- ○ [PEP];[FDP] ≅ 5.0 · 10⁻⁴ mole ·l⁻¹ : $K_{1/2}$ = 0.6 · 10⁻⁴ mole ·l⁻¹ ; n_H = 0.7
- • [ADP];[FDP] ≅ 0.0 · 10⁻⁴ mole ·l⁻¹ : $K_{1/2}$ = 1.8 · 10⁻⁴ mole ·l⁻¹ ; n_H = 1.3
- • [FDP];[PEP] ≅ 5.0 · 10⁻⁵ mole ·l⁻¹ : $K_{1/2}$ = 6.6 · 10⁻⁷ mole ·l⁻¹ ; n_H = 0.6

FIG. 7. Kinetic properties of PK in hemolysate of patient S. W. with deficiency type B. Hemolysate (H) after Sephadex G-25 filtration (for conditions see Fig. 3); electrophoresis and Hill plots (for assay conditions, see Fig. 3).

as compared to the properties of PK from normal erythrocytes. The same was true with respect to the transformation of these curves into hyperbolic forms by saturating the enzyme with FDP (PEP varied) or PEP (ADP varied), lowering $K_{\frac{1}{2}}$ and n_H. The absolutely smaller values for $K_{\frac{1}{2}}$ and n_H are explained by a different relation of the fractions A, B and C in favor of the latter in this patient. The $K_{\frac{1}{2}}$ values for PEP and ADP were normal. The striking difference, however, is the $K_{\frac{1}{2}}$ value for FDP, which was tenfold higher than normal in the unfractionated hemolysate. The heterozygous parents of this patient showed values 1.7- and 2.5- fold higher than normal.

Fractions of PK obtained by ammonium sulfate precipitation of hemolysate of the B phenotype did not show significant differences, as compared to the same fractions obtained from normal hemolysate, with respect to $K_{\frac{1}{2}}$ and n_H for PEP and ADP (Fig. 8, 9). However, a major difference was found in FDP enzyme activation. The concentration of FDP required for half maximal enzyme activation with PK fractions A and B was much higher than normal. The $K_{\frac{1}{2}}$ values were increased 100 times the normal.

PK obtained from type A-deficient erythrocytes. In contrast to type B the kinetic properties of the abnormal PK in type A are nearly normal. The $K_{\frac{1}{2}}$ of FDP activation is elevated only to a minor extent (Table 1). How-

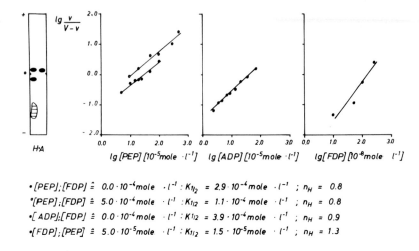

- $[PEP];[FDP] \cong 0.0 \cdot 10^{-4}$ mole $\cdot l^{-1}$: $K_{1/2}$ = 2.9 $\cdot 10^{-4}$ mole $\cdot l^{-1}$; n_H = 0.8
- $[PEP];[FDP] \cong 5.0 \cdot 10^{-4}$ mole $\cdot l^{-1}$: $K_{1/2}$ = 1.1 $\cdot 10^{-4}$ mole $\cdot l^{-1}$; n_H = 0.8
- $[ADP];[FDP] \cong 0.0 \cdot 10^{-4}$ mole $\cdot l^{-1}$: $K_{1/2}$ = 3.9 $\cdot 10^{-4}$ mole $\cdot l^{-1}$; n_H = 0.9
- $[FDP];[PEP] \cong 5.0 \cdot 10^{-5}$ mole $\cdot l^{-1}$: $K_{1/2}$ = 1.5 $\cdot 10^{-5}$ mole $\cdot l^{-1}$; n_H = 1.3

FIG. 8. Kinetic properties of PK fraction A isolated from type B deficiency (patient S.W.). Sephadex G-100 filtration of hemolysate (for conditions, see Fig. 4). $(NH_4)_2SO_4$ precipitation at 0.20 to 0.38 saturation (A); electrophoresis and Hill plots (for assay conditions, see Fig. 3).

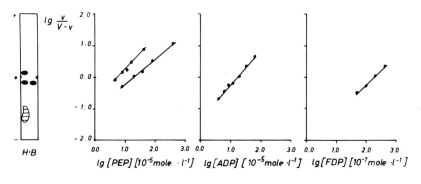

- $[PEP];[FDP] \cong 0.0 \cdot 10^{-4}$ mole $\cdot l^{-1}$: $K_{1/2}$ = 1.9 $\cdot 10^{-4}$ mole $\cdot l^{-1}$; n_H = 0.8
- $[PEP];[FDP] \cong 5.0 \cdot 10^{-4}$ mole $\cdot l^{-1}$: $K_{1/2}$ = 0.5 $\cdot 10^{-4}$ mole $\cdot l^{-1}$; n_H = 1.0
- $[ADP];[FDP] \cong 0.0 \cdot 10^{-4}$ mole $\cdot l^{-1}$: $K_{1/2}$ = 1.7 $\cdot 10^{-4}$ mole $\cdot l^{-1}$; n_H = 1.1
- $[FDP];[PEP] \cong 5.0 \cdot 10^{-5}$ mole $\cdot l^{-1}$: $K_{1/2}$ = 1.8 $\cdot 10^{-5}$ mole $\cdot l^{-1}$; n_H = 0.9

FIG. 9. Kinetic properties of PK fraction B isolated from type B deficiency (patient S.W.). Sephadex G-100 filtration of hemolysate (for conditions, see Fig. 4). $(NH_4)_2SO_4$ precipitation at 0.38 to 0.45 saturation (B); electrophoresis and Hill plots (for assay conditions, see Fig. 3).

TABLE 1. *Comparison of FDP activation of erythrocyte PK in normal individuals and cases with PK deficiency types A and B (assay conditions as in Fig. 3).*

	Normal enzyme		PK (–) type A patient R.Z.		PK (–) type B patient S.W.	
	$K_{\frac{1}{2}}$	n_H	$K_{\frac{1}{2}}$	n_H	$K_{\frac{1}{2}}$	n_H
Hemolysate	0.6×10^{-7}M	1.7	– – –	– –[a]	6.6×10^{-7}M	0.6
$(NH_4)_2SO_4$–precipitations						
A) 0.20 to 0.38	0.6×10^{-7}M	1.5	3.6×10^{-7}M	1.0	1.5×10^{-5}M	1.3
B) 0.38 to 0.45	1.7×10^{-7}M	1.6	4.6×10^{-7}M	1.2	1.8×10^{-5}M	0.9
C) 0.45 to 0.70	1.1×10^{-7}M	1.5	5.0×10^{-7}M	0.8	5.0×10^{-7}M	1.3
Average values of n_H		1.5		1.0		1.3

[a] Valves too small for reliable estimation.

ever, the average Hill exponent value of the ammonium sulfate fractions is only 1.0, whereas it is normally 1.5.

A comparison of the different $K_{\frac{1}{2}}$ and n_H values for FDP in the ammonium sulfate fractions obtained from hemolysates of normal and PK-deficient erythrocytes is given in Fig. 10. In normal erythrocytes the mean value of half maximal activation with FDP is 1.0×10^{-7} M. The mean value of the Hill exponents combining the n_H value with the percentage of activity of the single fractions in comparison to the whole activity of A, B and C respectively is 1.5, indicating, in classical terms, two binding sites for FDP.

In erythrocytes of type A deficiency, the PK shows only small differences from normal for the mean value of $K_{\frac{1}{2}}$, but has a Hill exponent approaching 1.0. This indicates only one binding site for FDP in contrast to two binding sites of the normal enzyme. The PK fractions A and B from erythrocytes of type B deficiency require much higher concentrations of FDP for half maximal activation as compared to the same fractions isolated from normal erythrocytes. Fraction C is normal in this respect. The mean value of the Hill exponent is 1.3 indicating more than one binding site.

In summary, normal erythrocytes and types A and B PK-deficient erythrocytes are quantitatively characterized by different overall PK activities. A major qualitative difference between these three phenotypes is manifested in the FDP concentrations required for half maximal enzyme activation and in the values of FDP Hill exponents.

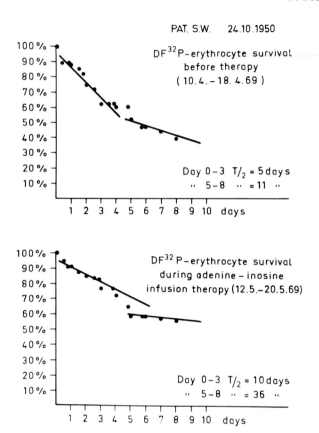

FIG. 10. Life span of red cells of patient S.W. (PK deficiency type B) before and during therapy with inosine, adenine and guanosine. One mmole inosine, and 0.05 mmole of adenine and guanosine in 500 ml of 0.9% NaCl solution was administered i. v. over a period of 3 hr daily.

Metabolic studies with normal and PK-deficient erythrocytes and their therapeutic application. Metabolic studies on normal and PK-deficient red cells (16–18) have demonstrated that by using adenine and inosine as glycolytic substrates, lactate production was about doubled in both, with a concomitant rise of the cellular ATP content. Adenine and inosine administration *in vivo* and *in vitro* resulted in an accumulation of FDP and triosephosphates in both types of cells. In addition, an elevation of PEP and 3-phosphoglycerate occurred in PK-deficient red cells only (Table 2). Encouraged by these results, we have studied the effect of *in vivo* administration of adenine and inosine in patients with PK-deficient

TABLE 2. *Glycolytic intermediates[a] in erythrocytes of patient R.Z. (PK deficiency type A) before, during and after infusion of 500 ml of isotonic saline, containing 0.22 M glucose, 1 mM adenine, 20 mM inosine, and 0.5 mM guanosine (18)*

	0	90 min	180 min	24 hr	Normal values	
	Start	Infusion	End		Mean	± SD
G-6-P	36	66	38	42	21	6
F-6-P	10	38	15	16	8	4
FDP	16	22	61	32	15	5
TP	21	70	127	35	47	20
3-PG	155	194	292	151	72	19
2-PG	17	29	37	25	30	10
PEP	50	79	112	71	21	6
Pyruvate[b]	64	52	36	53	134	47
Lactate[b]	518	1810	503	787	1450	470
L/P	8.1	34.8	14	14.8	11.3	2.7
ATP	843	980	935	1240	1280	126
ADP	292	268	244	381	279	59
AMP	61	58	62	83	51	21
ATP/ADP	2.9	3.7	3.8	3.3	4.9	1
ATP + ADP + AMP	1196	1306	1244	1704	1626	174

[a] nmole/ml erythrocytes
[b] nmole/ml blood

hemolytic anemia. Daily i.v. administration of adenine, inosine and guanosine in type B PK deficiency over a period of three weeks was followed by a rise of the hemoglobin level from 7 to 10 g/100 ml, a drop of reticulocyte content from 60 to 30% and by a prolongation of the red cell survival time (Fig. 10). Clinically, a marked improvement in the general condition of the patients was noted. Similar results were not observed in a case of type A PK-deficient hemolytic anemia.

The question arises as to why this therapeutic response was observed only in type B deficiency. In view of the kinetic results discussed, the possible effect of an elevated red cell FDP content on glycolysis is especially interesting. In both types of deficiency, PEP activates PK; however in type B, FDP activation may be of additional importance. Although we do not know the exact $K_{\frac{1}{2}}$ of FDP activation of PK within the intact red cell of a type B PK-deficient patient, it is probably of the order of magnitude of the intracellular FDP concentration observed. This is normally about 2.0×10^{-5} M.

In our experimental conditions, the $K_{\frac{1}{2}}$ for two of three fractions of the PK of patient S.W. was found to be 1.5 to 1.8×10^{-5} M, amounting to about 60% of total enzyme activity. On the contrary, in type A the affinity of PK to FDP is much higher and nearly normal, the $K_{\frac{1}{2}}$ of FDP activation being 3 to 5×10^{-7} M. This means, in any event, a value most probably far below the intracellular FDP content. If this value also represents *in vivo* conditions, it can be assumed that normal or type A PK is always highly saturated with FDP and that FDP activation plays no role in enzyme activity regulation. In type B PK deficiency, however, it seems reasonable to assume that not only the elevated PEP, but also the raised FDP concentration, resulting from adenine-inosine treatment, increases the PK-reaction velocity, thus affecting the cell metabolism and survival.

Our thanks to Mrs. Krey for technical assistance and to Dr. Sh. Moses for his help in translating this paper.

Supported by grants from Deutsche Forschungs-Gemeinschaft and Stiftung Volkswagenwerk.

REFERENCES

1. WALLER, H. D. and LÖHR, G. W. Hereditary nonspherocytic enzymopenic hemolytic anemia with pyruvate kinase deficiency. *Proc. IX Cong. int. Soc. Haemat. Mexico City*, 1962, p. 257.
2. PAGLIA, D. E., VALENTINE, W. N., BAUGHAM, M. A., MILLER, D. R., REED, C. F. and MCINTYRE, O. R. An inherited molecular lesion of erythrocyte pyruvate kinase. Identification of a kinetically aberrant isozyme associated with premature hemolysis. *J. clin. Invest.* **47**: 1929, 1968.
3. BOIVIN, P. and GALAND, C. Recherche d'une anomalie moléculaire lors des déficits en pyruvate kinase érythrocytiare. *Nouv. Rev. franc. Hémat.* **8**: 201, 1968.
4. SACHS, J. R., WICKER, D. J., GILCHER, R. O., CONRAD, M. E. and COHEN, R. J. Familial hemolytic anemia resulting from an abnormal red blood cell pyruvate kinase. *J. Lab. clin. Med.* **72**: 359, 1968.
5. FUSCO, F. A., BUSCH, D., NEGRINI, A. C. and AZZOLINI, A. Anemia emolitica congenita non sferocitica da anomalia della piruvato chinasi. *Haematologica* **11**: 836, 1966.
6. WIESMANN, U., TÖNZ, O., RICHTERRICH, R. and VERGER, P. Die Erythrozyten-Pyruvatkinase bei Gesunden und bei nichtsphärozytärer hämolytischer Pyruvatkinase-Mangel-Anämie. *Klin. Wschr.* **43**: 1311, 1965.
7. BUSCH, D. and PELZ, K. Zur Heterogenität des Pyruvatkinasemangels. *9th annual meeting of the Deutsche Gesellschaft fur Anthropologie. Homo* (Suppl.) 1967, p. 47.
8. BUSCH, D. and PELZ, K. Erythrozytenisolierung aus Blut mit Baumwolle. *Klin. Wschr.* **44**: 983, 1966.
9. BLUME, K. G., HOFFBAUER, R. W., LOHR, G. W. and RUDIGER, H. W. Genetische und biochemische Aspekte der Pyruvatkinase menschlicher Erythrozyten. *75th annual meeting of the Deutsche Gesellschaft für Innere Medizin, Wiesbaden*, 1969.
10. FELLENBERG, R. V., RICHTERRICH, R. and AEBI, H. Elektrophoretisch verschieden wandernde Pyruvatkinase aus einigen Organen der Ratte. *Enzym. Biol. clin.* (*Basel*) **3**: 240, 1963.
11. HESS, B., HAECKEL, R. and BRAND, K. FDP-activation of yeast pyruvate kinase. *Biochem. biophys. Res. Commun.* **24**: 824, 1966.

12. HAECKEL, R., HESS, B., LAUTERBORN, W. and WUESTER, K. H. Purification and allosteric properties of yeast pyruvate kinase. *Hoppe-Seylers Z. physiol. Chem.* **349**: 699, 1968.
13. KOLER, R. D. and VANBELLINGHEN, P. The mechanism of precursor modulation of human pyruvate kinase I by fructose diphosphate, *Advanc. Enzym, Regulat.* **6**: 127, 1968.
14. AEBI, H., SCHNEIDER, C. H., GANG, H. and WIESMANN, U. Separation of catalase and other red cell enzymes from hemoglobin by gel filtration. *Experientia (Basel)* **20**: 103, 1964.
15. BIGLEY, R. H., STENZEL, P., JONES, R. T., CAMPOS, J. O. and KOLER, R. D. Tissue distribution of human pyruvate kinase isozymes. *Enzym. Biol. clin. (Basel)* **9**: 10, 1968.
16. BUSCH, D. Congenitale nichtsphärocytäre hämolytische Anämie mit Mangel an erythrozytärer Pyruvatkinase. *Folia haemat.* (N.F.), *(Frankfurt)* **9**: 89, 1964.
17. BUSCH, D. Probleme des Erythrozytenstoffwechsels bei Anämien mit Pyruvatkinasemangel. *Folia haemat. (Lpz.)* **83**: 395, 1965.
18. BUSCH, D. und BOIE, K. Beiträge zum Problem des Pyruvatkinasemangels. *Folia haemat. (Lpz.)* **91**: 77, 1969.

NEW GLUCOSE-6-PHOSPHATE DEHYDROGENASE VARIANTS IN ISRAEL*

Association with Congenital Nonspherocytic Hemolytic Disease

BRACHA RAMOT, ISAAC BEN-BASSAT and MORDECHAI SHCHORY

Department of Hematology and WHO Regional Reference Center for Glucose-6-Phosphate Dehydrogenase, Government Hospital, Tel-Hashomer and Tel Aviv University Medical School, Israel

It has been well established that there is an association between congenital nonspherocytic hemolytic disease (CNHD) and sporadic glucose-6-phosphate dehydrogenase (G6PD) variants (1-11, Busch and Stubler, in preparation and Tanka and Beutler, unpublished). About 15 distinct G6PD variants have so far been found in patients with CNHD, while the common Mediterranean variant has also been found in a number of them (9,12).

During the past two and one half years, we have studied eight patients with CNHD and have characterized new variants of G6PD in three of them. In the other five patients, the common Mediterranean variant was found; while a ninth patient, who exhibited only one hemolytic episode during infectious hepatitis, was found to have yet another new variant. In this report we describe the properties of these new variants.

MATERIALS AND METHODS

Using the recommended procedures of the World Health Organization G6PD Standardization Committee, red cell G6PD was partially purified and characterized (13). The partially purified enzyme was subjected to electrophoresis in a Tris-EDTA-borate (TEB) system, pH 8.5, as described by Chernoff et al. (14), using Electrostarch (Madison, Wis). In one case

---

* Part of this study has been published in the *J. Lab. clin. Med.* **74**: 895, 1969.

phosphate buffer, pH 7.0, was also used (15). Using the method of Kirkman (16), the pH optimum curves were determined with buffers ranging from pH 5.5 to 11. Bovine albumin, in a final concentration of 1 mg/ml, was added to the partially purified diluted enzyme in order to determine thermostability. The purification and characterization of all enzymes was carried out at least twice.

Patient material. All eight patients were males, aged 20 to 30, with marked G6PD deficiency, and were referred to us with a history of multiple hemolytic episodes during drug administration or infection or both, as well as chronic indirect hyperbilirubinemia. Three of the patients were Jews of Iraqi origin (Cases 1, 2, 4*), three were Jews of North African origin (Cases 3, 6, 7), Case 8 was a Persian Jew and Case 5 was an Ashkenazi Jew. During the period of study, all cases had a normal Hb level, a normal or increased reticulocyte count and slight or marked indirect hyperbilirubinemia. In Cases 3 and 7, autologous Cr^{51} labeled red cell survival time was shortened ($T^1/_2$ = 12 and 20 days respectively). The spleen was not palpable in any of the cases, and all additional parameters, such as osmotic fragility, autohemolysis, Coombs' test, and Hb electrophoresis were normal. An additional new variant of G6PD was found in a ninth patient, an Iraqi Jew, who had had a hemolytic episode during the course of infectious hepatitis, but showed no signs of hemolysis on follow-up.

RESULTS

Fig. 1 to 5 represent the characteristics of the different enzymes in the patients. The normal ranges given in the figures were obtained from about 60 control subjects. While it is clear that the enzymes from five of the patients had the same characteristics as the Gd Mediterranean variant, the enzymes from the other four patients showed distinct differences. Three brothers of patient 3 were also studied and although none of them showed an increased bilirubin level, their enzyme characteristics were similar to those of their sibling.

Distinct differences were found between this variant and the Gd Mediterranean variant with regard to mobility in TEB buffer, but no differences were found between the two variants on starch gel electrophoresis using phosphate buffer.

---

* Case 4 was published previously as an example of hemolysis due to aspirin administration (17).

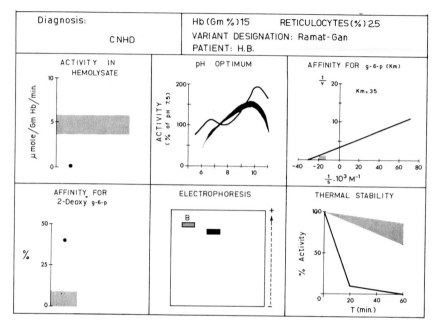

FIG. 1. Clinical and enzymatic characteristics of G6PD mutant, Gd Ramat-Gan (Case 1). Shaded areas indicate the normal range. B under "electrophoresis" indicates the position of the common GdB variant.

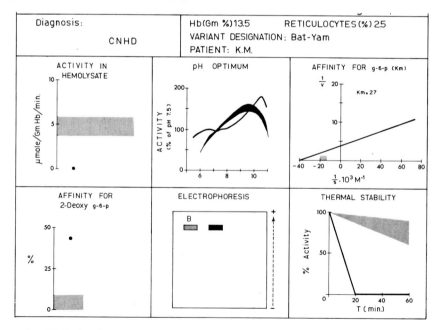

FIG. 2. Clinical and enzymatic characteristics of G6PD mutant, Gd Bat-Yam (Case 2).

New G6PD Variants in Israel

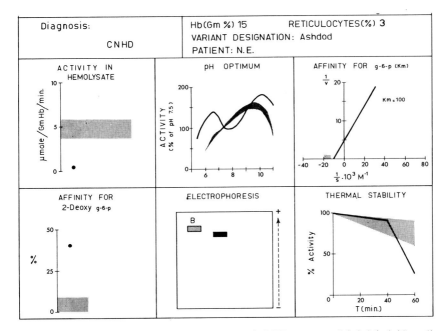

FIG. 3. Clinical and enzymatic characteristics of G6PD mutant, Gd Ashdod (Case 3).

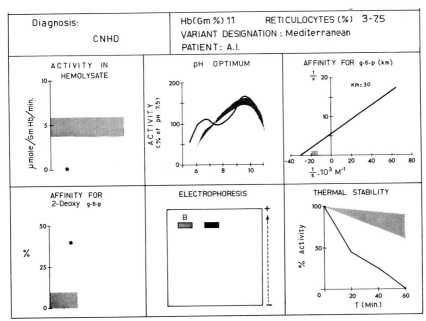

FIG. 4. Clinical and enzymatic characteristics of G6PD mutant, Gd Mediterranean (Case 4). The same enzymatic characteristics were found in Cases 5, 7 and 8.

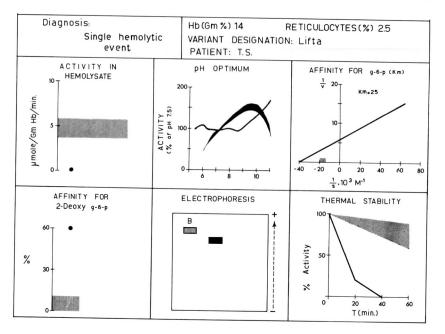

FIG. 5. Clinical and enzymatic characteristics of G6PD mutant, Gd Lifta (Case 6).

DISCUSSION

Since the electrophoretic and biochemical characteristics of the erythrocyte G6PD variants in three of our patients with CNHD seem to be different from those previously reported, we suggest they be called Gd Ramat-Gan, Gd Bat-Yam and Gd Ashdod. Likewise, we detected another unusual variant, Gd Lifta, in a patient who had had one hemolytic episode during infectious hepatitis (18).

Gd Bat-Yam differs from Gd Mediterranean by its marked lability. After being incubated for 20 min at 45 C, the enzyme becomes inactive. Gd Bat-Yam also differs from Gd Mediterranean in the pH curve (Fig. 2). Increased lability and slower electrophoretic mobility (92%) are the features that distinguish Gd Ramat-Gan from Gd Mediterranean. Gd Ashdod differs from Gd Mediterranean in its slower electrophoretic mobility, high affinity for 2-deoxyglucose-6-phosphate, and slightly increased Michaelis constant (K_m) (100 μM) for glucose-6-phosphate (Fig. 3).

The severity of the disease, manifested in all patients as a well established hemolytic syndrome, was found to have no correlation with the type of

G6PD variant. The patient who had the most marked hemolytic process, mostly from undetectable causes but probably due to mild viral infections, was one of the two patients with the Mediterranean G6PD variant. The patient with the Gd Ashdod variant exhibited signs of chronic hemolysis with hyperbilirubinemia, while his three siblings who had the same variant were completely normal.

Two features seem to be common to all five of the variants described in patients with CNHD: 1) A low affinity for glucose-6-phosphate manifested by high K_m as previously described in other variants: Freiburg (Busch and Stubler, in preparation), Albuquerque (9), Oklahoma (1), Clichy (10), Paris (10), Beaujon (10) and Milwaukee (5). This phenomenon was also observed in the Gd Ashdod variant. 2) A high *in vitro* lability of the enzyme observed in the Gd Ramat-Gan variant and the Gd Bat-Yam variant and previously reported in other variants: Chicago (3), Duarte (9), Albuquerque (9), Bangkok (11), Oklahoma (1), Paris (10), Ohio (4) and Eyssen (2).

Possibly because of these two features the small amounts of the enzyme present in the red cell are functionally less efficient than the Gd Mediterranean enzyme and this may explain the clinical differences. However, the fact that the three siblings of patient 3 exhibited no hemolysis seems to suggest that the variant present is not the sole determinant of the chronic hemolytic process.

Various possibilities have been suggested by Beutler et al. (9) to explain the association between the clinical syndrome and the Gd Mediterranean variant as exemplified by two of our patients. The most likely explanation seems to be that CNHD results from the interaction of Gd Mediterranean with another autosomal gene. There is a possibility that these two cases of defects in a population with high incidence of G6PD deficiency is coincidental, as seen in nonAshkenazi Jews.

It is very important to note that one of the patients with the Gd Mediterranean variant is an Ashkenazi Jew. This is an ethnic group which has a very low (0.2% or less) incidence of G6PD deficiency.

Additional studies of patients with CNHD and their families are needed in order to establish a relationship between the clinical syndrome and the G6PD variants.

REFERENCES

1. KIRKMAN, H. N. and RILEY, H. D., Jr. Congenital nonspherocytic hemolytic anemia. *Amer. J. Dis. Child.* **102**: 313, 1961.
2. BOYER, S. H., PORTER, I. H. and WEILBACHER, R. G. Electrophoretic heterogeneity of glucose-6-phosphate dehydrogenase and its relationship to enzyme deficiency in man. *Proc. nat. Acad. Sci. (Wash.)* **48**: 1868, 1962.
3. KIRKMAN, H. N., ROSENTHAL, I. M., SIMON, E. R., CARSON, P. E. and BRINSON, A. G. "Chicago I" variant of glucose-6-phosphate dehydrogenase in congenital hemolytic disease. *J. Lab. clin. Med.* **63**: 715, 1964.
4. PINTO, P. V. C., NEWTON, W. A., Jr. and RICHARDSON, K. E. Evidence for four types of erythrocyte glucose-6-phosphate dehydrogenase from G-6-PD deficient human subjects. *J. clin. Invest.* **45**: 823, 1966.
5. WESTRING, D. W. and PISCIOTTA, A. V. Anemia, cataracts, and seizure in patient with glucose-6-phosphate dehydrogenase deficiency. *Arch. intern. Med.* **118**: 385, 1966.
6. WALLER, H. D., LOHR, G. W. and GAYER, J. Hereditäre nichtsphärocytäre hämolytische Anämie durch Glucose-6-Phosphatdehydrogenase-Mangel. (Bildung eines Enzymproteins mit veränderten Eigenschaften in den Blutzellen einer deutschen Familie.) *Klin. Wschr.* **44**: 122, 1966.
7. HELGE, H. and BORNER, K. Kongenitale nichtsphärozytäre hämolytische Anämie, Katarakt und Glucose-6-Phosphat-Dehydrogenase-Mangel. *Dsch. med. Wschr.* **91**: 1584, 1966.
8. WEINREICH, J., BUSCH, D., GOTTSTEIN, U., SCHAEFER, J. and ROHR, J. Über zwei neue Fälle von hereditärer nichtsphärocytärer hämolytischer Anämie bei Glucose-6-Phosphatdehydrogenase-Defekt in einer norddeutschen Familie. *Klin. Wschr.* **46**: 146, 1968.
9. BEUTLER, E., MATHAI, C. K. and SMITH, J. E. Biochemical variants of glucose-6-phosphate dehydrogenase giving rise to congenital nonspherocytic hemolytic disease. *Blood* **31**: 131, 1968.
10. BOIVIN, P. and GALAND, C. Nouvelles variantes de la glucose -6-phosphate dehydrogenase erythrocytaire. *Rev. Franç. Étud. clin. biol.* **13**: 30, 1968.
11. TALALAK, P. and BEUTLER, E. G-6-PD Bangkok. A new variant found in congenital nonspherocytic hemolytic disease (CNHD). *Blood* **33**: 5, 1969.
12. SCHETTINI, F. and MELONI, T. Characterization of glucose-6-phosphate dehydrogenase in Sardinian children with congenital nonspherocytic haemolytic anaemia. *Acta haemat. (Basel)* **37**: 198, 1967.
13. Standardization of procedures for study of G-6-PD. Report of a WHO scientific group. *Wld Hlth Org. techn. Rep. Ser.* **366**: 1967.
14. CHERNOFF, A., PETIT, N. and NORTHRUP, I. The amino acid composition of hemoglobin. The preparation of purified hemoglobin fractions by chromatography on cellulose exchangers and their identification by starch-gel electrophoresis using Tris-borate EDTA buffer. *Blood* **25**: 646, 1965.
15. MATHAI, C. K., OHNO SUSUMU and BEUTLER, E. Sex-linkage of the glucose-6-phosphate dehydrogenase gene in Equidae. *Nature (Lond.)* **210**: 115, 1966.
16. KIRKMAN, H. N. Glucose-6-phosphate dehydrogenase from human erythrocytes. I. Further purification and characterization. *J. biol. Chem.* **237**: 2364, 1962.
17. SZEINBERG, A., KELLERMAN, J., ADAM, A., SHEBA, Ch. and RAMOT, B. Haemolytic jaundice following aspirin administration to a patient with a deficiency of glucose-6-phosphate dehydrogenase in erythrocytes. *Acta haemat. (Basel)* **23**: 58, 1960.
18. RAMOT, B., BEN-BASSAT, I. and SHCHORY, M. New glucose-6-phosphate dehydrogenase variants observed in Israel. *J. Lab. clin. Med.* **74**: 895, 1969.

GLYCOGEN METABOLISM AND GLYCOLYSIS IN ERYTHROCYTES FROM PATIENTS WITH GLYCOGEN STORAGE DISEASE TYPE III AND NORMAL SUBJECTS

S. W. MOSES, N. BASHAN and R. CHAYOTH

Department of Pediatric Research, Central Negev Hospital and Negev University of Beersheba, Israel

INTRODUCTION

Mature RBC have, in contrast to other cells of the human body, limited metabolic capacities. The absence of nuclei, mitochondria and ribosomes excludes DNA directed protein synthesis, oxidative phosphorylation and a functional citric acid cycle. As a result of these limitations, only a small number of substrates can be utilized by the mature RBC, and metabolic control mechanisms do not operate by adapting enzyme concentration to metabolic needs. Mature human RBC meet their energy requirements primarily from the metabolism of carbohydrates, chiefly glucose, although other monosaccharides can be utilized. Normally 6 to 8 μmole of glucose/g Hb per hr are broken down to pyruvate and lactate, via the well known Embden-Meyerhoff pathway. Net generation of ATP and the cycle of nicotinamide adenine dinucleotide (NAD)-reduced nicotinamide adenine dinucleotide (NADH) ensues. In addition, a small fraction of the glucose-6-phosphate formed is metabolized via the oxidative hexosemonophosphate pathway yielding reduced nicotinamide adenine dinucleotide phosphate (NADPH). Another interesting side reaction, possibly of major importance, results in the accumulation of uniquely large amounts of 2,3-diphosphoglycerate (2,3-DPG) in the Rapaport-Lübering cycle (1).

In spite of their dependence on glucose for energy production, normal RBC have virtually no carbohydrate stores. However, in glycogen storage disease (GSD), Type III, abnormally high glycogen concentrations are found in the RBC (2). The availability of erythrocytes from GSD Type III patients permitted a comparative study of normal and glycogen-rich cells.

MATERIALS AND METHODS

The subjects of this study were 17 patients in whom the clinical and biochemical enzymatic diagnosis of GSD Type III had been made. None of them had abnormal Hb values or significant elevations of reticulocyte counts or plasma bilirubin levels. In some of the GSD patients, RBC life span was determined by Cr^{51} labeling, osmotic fragility and density distribution (3), all of which were found to be within normal limits. Normal control blood was obtained from hospital personnel.

Preparation of erythrocytes. The blood samples were drawn into heparinized tubes and the RBC sedimented by centrifugation in the cold. After aspiration of the supernatant plasma with the buffy coat, the remaining blood was pressed through a cotton sieve, as described by Busch (4). With this technique, virtually all thrombocytes and leukocytes were removed. The RBC were subsequently washed three times with 10 volumes of cold 0.9% NaCl before use.

Analytical methods. Hemoglobin was measured as cyanmethemoglobin (5). Deproteinization was performed with Somogyi reagents (6) for glucose determination, with 6% trichloroacetic acid for analysis of lactate, and with 5% perchloric acid for glycolytic intermediates and adenine nucleotides. Glucose was determined by the glucose oxidase method (7) and lactate by the enzymatic method of Horn and Burns (8).

The activity of hexokinase, lactate dehydrogenase, phosphoglucomutase, glucose-6-phosphate dehydrogenase, 6-phosphogluconate dehydrogenase and pyruvate kinase were measured by methods previously described (9–14). Amylo-1,6-glucosidase and phosphorylase activities were determined according to the procedure of Hers (15) and that of uridine diphosphoglucose (UDPG) glycogen-glucosyltransferase by the method described by Cornblath et al. (16). 1,4-glucan:1,4-glycosyl transferase activity was determined by the method of Larner (17).

To remove hemoglobin, hemolysates were suspended in DEAE cellulose prepared with 3 mM buffer phosphate, pH 7.0, after which the mixture was rinsed in a similar phosphate buffer. The enzyme, hexokinase, was finally eluted with 0.5 M KCl.

Glycolytic intermediates and adenine nucleotides were measured as follows: 2,3-DPG was determined by the Krimsky assay (18); phosphate concentration was determined by the Fiske-Subbarow method (19); ATP content was measured by the glyceraldehyde-3-phosphate dehydrogenase and phosphoglycerokinase assay system of Adam (20); ADP was determined by rephosphorylation in the presence of pyruvate kinase and lactate dehydrogenase; AMP was measured in a similar system by prior addition of ATP and myokinase, acidification and determination as ADP (21). Triosephosphates and fructose diphosphate were determined enzymatically utilizing aldolase, triosephosphate isomerase and glycerol dehydrogenase, as described by Bucher and Hohorst (22).

The changes in reduced pyridine nucleotide, upon which most of these assays depend, were measured spectrophotometrically or fluorometrically, depending on the concentration of the investigated metabolite.

Incubation conditions. The production of $C^{14}O_2$ from 1-C^{14}-glucose was studied by incubating a 30% RBC suspension in a reaction mixture (final vol 4 ml) containing 4 mM glucose, 0.15 μc 1-C^{14}-glucose, 2.5 mM Na_2HPO_4, 0.12 M glycyl-glycine buffer, pH 7.8, in sealed vials equipped with center wells containing 0.5 ml of 15% KOH absorbed on No. 1 Whatman filter paper. At the end of the incubation, 0.4 ml of 1 N H_2SO_4 was injected into the medium and the vial shaken for one additional hr. The $C^{14}O_2$ absorbed on the filter paper was subsequently counted in a liquid scintillation counter.

For glucose utilization and lactate formation, a 10% RBC suspension was used in the same incubation mixture as described above. Aliquots of the incubation mixture were assayed each hr and the values given represent mean hourly changes.

Hemolysates for the measurement of glycolytic rates and U-C^{14}-glucose incorporation studies were prepared by the addition of equal volumes of H_2O to packed RBC. One volume of the hemolysate was incubated with 1 vol of an incubation mixture containing 2 mM glucose, 2.5 μc U-C^{14}-glucose, 0.75 mM ATP, 0.75 mM NADP, 0.3 mM $MgCl_2$, 0.5 mM K_2HPO_4, 0.1 M glycyl-glycine buffer, pH 7. 4.

Glucose incorporation into glycogen and assay of glycogen. Studies on U-C^{14}-glucose incorporation into glycogen were performed on intact RBC prepared as described above. Incubation conditions were as previously described except that 2.5 μc/vial of U-C^{14}-glucose was used. After the incubation, nonlabeled carrier glycogen was added, and cold trichloroacetic acid added as described by Sidbury et al. (2).

Incubations without erythrocytes did not result in any nonspecific glucose-glycogen exchange. Analysis of the distribution of the U-C^{14}-glucose in the outer branches and inner core of the glycogen was carried out after β-amylolysis (23). The maltose liberated by β-amylolysis was separated from other compounds by descending paper chromatography utilizing butanol-pyridine-water 3:2.5:1.5 (v/v/v). Maltose was determined by the

Somogyi-Nelson method (24). Radioactivity of the remaining limit dextrin and of the separated maltose was measured in a liquid scintillation counter.

Glycogen was recovered from the RBC after repeated precipitation of the protein with cold 6% trichloroacetic acid, collection of the supernatant fluid and precipitation of glycogen in cold 95% ethanol after the addition of 0.18% Na_2SO_4. Glycogen was determined with a diazyme reagent (Miles Chemicals, Elkhard Ind.) (25).

Separation of glycogen from heteropolysaccharides was achieved by column chromatography utilizing Bio-Gel-G-200 and water as the eluting phase (Keppler, D., personal communication).

RESULTS

Red cell glycogen concentration. When measured by spectrophotometric methods, no glycogen was detected in normal RBC. However, utilizing a highly sensitive radioactive method, a low concentration of 2.9 ± 0.5 μg glycogen/g Hb was found. Abnormally high glycogen concentrations were detected in GSD Type III erythrocytes measured by spectrophotometric

TABLE 1. *Activities of enzymes concerned with glycogen metabolism in RBC from normal and Type III GSD subjects*

Enzyme	Units	Activity		P
		Normal	Type III GSD	
UDPG glycogen glycosyl-transferase	μmole glucose incorporated into glycogen	1.79±0.16 (7)	1.95±0.24 (4)	NS
α-1,4 glucan: α-1,4 glucan-6-glycosyl-transferase (brancher enzyme)	Units[a]/g Hb	8.3 (3)	10.3 (3)	
Amylo-1, 6-glucosidase (debrancher enzyme)	% glucose incorporated/g Hb per hr	0.24±0.04(13)	0.004±0.002(8)	<0.005
Phosphorylase	μmole Pi_4 released/g Hb per hr	5.9 ±1.2 (9)	9.7±1.7 (8)	<0.001

[a] One unit equals change in optical density of 0.001/min at 520 mμ.
Values are means ± SE for the number of cases given in parentheses.
NS = not significant.

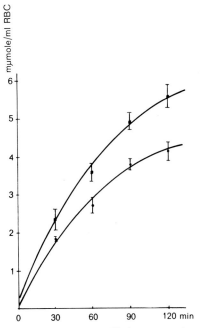

FIG. 1. Mean values are represented by ■ in the normal RBC and by ● in the RBC of Type III GSD patients. The vertical lines indicate SE of the mean.

methods. The glycogen concentration in 17 cases tested ranged from 200 to 3,000 µg/g Hb with a mean of 1,054 µg/g Hb.

Enzymes of glycogen metabolism. Activities of UDPG glycogen synthetase and of the branching enzyme were not significantly different in RBC from normal and GSD Type III subjects. As expected, virtually no debrancher activity was detected in RBC of any GSD Type III patient; however a significant increase in phosphorylase activity was disclosed in the latter cells (Table 1).

Glycogen synthesis from glucose incorporation of $U-C^{14}$-glucose into glycogen was increased during the first 2 hr in both normal and GSD Type III RBC in a similar manner, but at a slightly lower rate in the affected cells (Fig. 1). A peak of radioactivity was reached in both types of cells after 2 hr. Similar incorporation rates were observed in hemolysates. To confirm that the radioactive glucose was actually incorporated into glycogen in

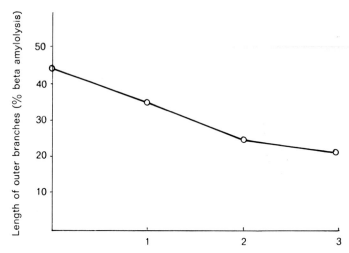

FIG. 2. Rate of glycogen breakdown in intact erythrocytes of GSD Type III: shortening of outer branches during incubation. In the incubation mixture glucose was omitted (see Methods).

TABLE 2. *Glycolytic rates in intact RBC of normal subjects and Type III GSD patients*

RBC	Glucose disappearance µmole/g Hb	Lactate formation µmole/g Hb
Normal	7.5 ± 0.34 (58)	12.8 ± 0.62 (26)
GSD Type III	10.9 ± 0.86 (10)	11.8 ± 0.82 (8)
P	< 0.001	NS

Values are means ± SE for the number of experiments in parentheses.
Aliquots of the incubation mixture (see Methods) were assayed hourly.

normal RBC, column chromatography was performed to separate hetero- and conjugated polysaccharides from glycogen, and the glycogen was isolated and subjected to β-amylolysis. After U-C^{14}-glucose incorporation, an equal volume of 6% $HgCl_2$ was added to the incubation mixture in order to precipitate the protein. Unlabeled glycogen carrier was added to the supernatant fluid which was then washed and precipitated with ethanol as described in Methods. The polysaccharide suspension was subsequently passed through a Bio-Gel G-200 column. The peaks of radioactivity in the eluate conformed with the peaks characteristic of glycogen (Keppler, D., personal communication) as measured by spectrophotometric methods.

β-amylolysis of the precipitated glycogen after U-C^{14}-glucose incorporation indicated that the radioactivity resided in expected proportions in the limit dextrin and maltose fractions.

Glycogen breakdown. Both normal and affected RBC, which had been incubated with U-C^{14}-glucose, were washed and reincubated with non-labeled glucose. This resulted in a decrease in glycogen radioactivity of 75.6% in normal cells and of 85.1% in GSD Type III RBC. Upon incubating glycogen rich cells in the absence of substrate and measuring the length of outer branches by β-amylolysis at various time intervals, a shortening of outer branches occurred over a 3-hr period (Fig. 2).

Glycolysis. The results shown in Table 2 indicate that the rate of glucose utilization by RBC from GSD Type III patients was significantly higher ($P < 0.001$) than the values obtained with normal RBC. However, the rate of lactate formation was similar in the two types of cells. Most of the glucose utilized was converted into lactate in normal cells (glucose/lactate ratio 0.85), whereas the excess glucose utilized in the RBC of GSD Type III patients could not be accounted for as lactate. The findings were similar in both intact cells and in hemolysates (Table 3). Production of C^{14}O$_2$ from 1-C^{14}-glucose was greater in GSD Type III cells when methylene blue was added as an electron acceptor (Table 4).

TABLE 3. *Glycolytic rates in hemolysates in normal and Type III GSD RBC*

RBC	Glucose disappearance µmole/g Hb	Lactate formation µmole/g Hb
Normal (5)	8.1 ± 0.43	9.15 ± 0.23
GSD Type III (5)	13.05 ± 0.61	9.9 ± 0.47

Values are means ± SE for the number of subjects given in parentheses.
Aliquots of the incubation mixture (see Methods) were assayed hourly.

TABLE 4. *C^{14}O$_2$ production from 1-C^{14}-glucose by RBC from normal and Type III GSD subjects*

	Normal µmole/g Hb per hr	Type III GSD µmole/g per hr	P
Control	0.074 ± 0.008 (8)	0.10 ± 0.012 (14)	NS
Methylene blue 0.01%	2.28 ± 0.26 (6)	3.60 ± 0.27 (12)	<0.01

Values are means ± SE for number of experiments in parentheses.
The cells were incubated for 1 hr.

TABLE 5. *Activity of glycolytic enzymes in RBC from normal and Type III GSD subjects*

Enzyme	Normal	Type III GSD	P
Hexokinase	13.4± 0.02 (4)	11.2± 0.14 (4)	NS
Phosphoglucomutase	67.0± 10.1 (4)	66.9± 7.35 (4)	NS
Pyruvate kinase	1104 ± 72.5 (4)	1002 ± 61.4 (4)	NS
Lactate dehydrogenase	4338 ±221 (4)	5070 ±197 (4)	NS
Glucose-6-phosphate dehydrogenase	238.0± 19.7 (6)	227.5± 15.4 (5)	NS
6-phosphogluconate dehydrogenase	61.0± 5.8 (5)	87.2± 6.3 (5)	<0.02

Values are expressed as µmole/g Hb per min ± SE for the number of subjects given in parentheses.

TABLE 6. *Concentration of 2,3-DPG during glycolysis in RBC from normal and Type III GSD patients*

Time	30 min µmole/g Hb	90 min µmole/g Hb	Change/hr µmole/g Hb
Patients (6)	11.8±0.5	20.6±0.73	8.8±0.83
Normal (14)	8.8±0.79	11.2±0.43	2.4±0.27

Values are means ± SE for the number of cases in parentheses.

Enzymes of glycolysis. Properties of partially purified hexokinase [Michaelis Constant (K_m) for glucose and pH activity curves] were not different in the two types of cells. No significant differences were found in the activities of several other enzymes of the glycolytic pathway, except that a significantly higher value for 6-phosphogluconate dehydrogenase was found in RBC of GSD Type III patients (Table 5).

Effect of glycogen on glycolysis. Addition of rabbit liver glycogen to hemolysates did not affect the glycolytic rate.

Concentration of glycolytic intermediates and cofactors. Concentration of fructose diphosphate and of triosephosphates were similar in normal and GSD Type III cells. However a fall of ADP and AMP was observed during glycolysis resulting in ADP/ATP ratios of 0.10 as compared to 0.45 in normals, indicating a higher rate of ATP regeneration. The levels of 2,3-DPG were strikingly higher in the affected as compared to normal cells (Table 6). This accumulation of 2,3-DPG could quantitatively account for glucose taken up but not recovered as lactate in the affected cells.

DISCUSSION

The studies presented in this report provide evidence that normal mature RBC maintain an active glycogen metabolism. The RBC glycogen determination employed in this report is based on an efficient isolation of erythrocytes and measurement of glycogen by sensitive radioactive incorporation methods. Previous studies, employing the conventional separation procedure, did not reveal an active glycogen metabolism in the mature red cell (1, 26 and Bartel, H., personal communication).

In order to present further evidence that the radioactive glucose was incorporated specifically into glycogen and not into other heteropolysaccharides or conjugated compounds, separation by column chromatography showed that the radioactivity resided in glycogen. Furthermore, exposure of the radioactive material to β-amylolysis resulted in the recovery of radioactive maltose. These two experiments provide additional evidence that the radioactive polysaccharide precipitated from RBC was true glycogen.

Although the minute amount of glycogen found in the normal RBC does not seem to play any major role in the energy metabolism, it was considered important to establish that an active glycogen metabolism does exist in both normal RBC and RBC of GSD Type III patients.

Sidbury et al. (2) reported abnormally high concentrations of glycogen in GSD Type III RBC. We confirmed their findings of increased glycogen concentration, mostly within 200 to 300 times of normal, but sometimes as much as 1,000 times the normal concentration. It is obvious that these abnormal accumulations of glycogen are related to the debrancher enzyme defect present in these cells. However, no studies had been performed to establish whether this accumulation represents an abnormal steady state of a metabolically active glycogen or an inactive vestigial remnant from an immature RBC stage, as postulated by Sidbury et al. Subsequent experiments were performed to establish whether an active glycogen metabolism could be shown to occur in mature RBC from normal subjects and from patients with GSD Type III. The enzymatic machinery for glycogen metabolism was present in both types of cells. However, as expected, no debrancher enzyme activity could be shown in GSD Type III cells. On the other hand, these cells had significantly higher phosphorylase activity than normal cells, which correlates well with the increased rate of glycogen breakdown observed in these cells. Active incorporation of glucose into glycogen was demonstrated in both types of cells, as shown in Fig. 1. The inital slope represents the net rate of incorporation, whereas in the normal cells, where the flow of glycosyl units in and out of glycogen is assumed to be unhindered, a peak reached after 2 hr represents minimal or subminimal values of glycogen concentrations. The concentration of glycogen in normal RBC derived from the radioactive methods is about one order of magnitude lower than the limit of sensitivity of glycogen determination by spectrophotometric methods. As different conditions exist in glycogen-rich, debrancher-deficient cells, the peak observed after 2 hr cannot be correlated with glycogen concentrations in these RBC. It is evident from the incorporation and breakdown studies that glycogen turnover in glycogen-rich cells occurs within minutes to hours, a time span that correlates quite well with glycogen turnover in the liver.

The evidence presented in this report allows the conclusion that both the normal and affected RBC maintain an active glycogen metabolism. The glucose is both incorporated into and released from glycogen. In glycogen-rich cells the outer branches are shortened during incubation in the absence of added glucose. Since leukocytes, thrombocytes and reticulocytes have been virtually removed from the erythrocyte preparations before use, this glycogen metabolism must be attributed to the mature human RBC. Thus, the concept that the mature RBC has no active glycogen metabolism is no longer tenable. This is not surprising in view of the demonstration that all the enzymes concerned with glycogen synthesis and breakdown are present in normal RBC.

As GSD Type III RBC contain abnormally high concentrations of a metabolically active glycogen, it was of interest to determine whether an abnormal control of glycolysis can be demonstrated in these unique cells. The rate of glucose utilization by RBC from patients suffering from GSD Type III was significantly higher than that seen in the normal RBC. However, less glucose was recovered as lactate, which in the normal RBC accounts for about 85% of the glucose utilized. Similar findings were observed in hemolysates (Tables 2, 3).

The possibility was raised that the increased glucose utilization in the glycogen-rich RBC is due in part to an increased oxidation of glucose along the pentose phosphate shunt. Production of $C^{14}O_2$ from $1-C^{14}$-glucose was significantly greater by cells from GSD Type III patients than by normal RBC when methylene blue was added as an electron acceptor.

The high rate of glycolysis observed in GSD Type III erythrocytes could also be a result of an abnormal activity of the enzyme regulating the initial phosphorylation of glucose. This possibility was unlikely in view of the normal K_m value for glucose and normal pH activity curve of hexokinase. However this does not entirely rule out the possibility of different hexokinase kinetics in view of recently published data by Garby and co-workers indicating that different binding of Hb by either 2,3-DPG or ATP can affect hexokinase activity in the intact cells (27).

No significant differences in the activities of other key enzymes of glycolysis were found except that 6-phosphogluconate dehydrogenase showed a significantly higher activity in all abnormal cases tested (Table 5). This enzyme has been considered to be rate controlling in the first part of the hexosemonophosphate shunt, provided sufficient NADP is regenerated in the presence of an electron acceptor (methylene blue). The higher activity of this part of the shunt, as shown by a higher rate of CO_2 production, correlated well with the demonstration of a higher 6-phosphogluconate dehydrogenase activity in RBC of patients with GSD Type III.

A higher glycolytic rate in the absence of reticulocytosis, a normal life span, and unchanged activities of several key glycolytic enzymes precluded the possibility that a younger cell population could be solely responsible for the higher rates of glucose utilization observed in the glycogen-rich cells. It is evident then that an abnormal control mechanism must operate in the affected cells, not reflected in maximal activities of the examined key enzymes of the Embden-Meyerhof pathway.

The possibility that glycogen per se stimulates glycolysis was also examined in hemolysates and the results were negative. It is undersood that structural differences in different types of glycogen may exist and, furthermore, that any specific localization of the glycogen within the erythrocyte cannot be imitated in hemolysates. Therefore only limited conclusions can be drawn from this experiment.

Inorganic phosphate is known to be a stimulator of glycolysis (28). Phosphate concentrations in whole blood and washed RBC were not different in the two types of cells; the higher glycolysis observed in affected RBC cannot, therefore, be attributed to a higher phosphate concentration.

In view of the findings that the higher glucose disappearance was not reflected in a higher lactate accumulation, it was of particular interest to study the concentrations of glycolytic intermediates during glycolysis. The most striking changes in this respect were observed in 2,3-DPG concentration. The concentration of 2,3-DPG is regulated by two enzymes of the phosphoglycerate shunt. 2,3-DPG mutase competes with the ADP-dependent phosphoglycerate kinase for a common substrate 1,3-DPG. A low ADP/ATP ratio, found in the glycogen-rich cells, favors 2,3-DPG formation (3). It is evident from Table 6 that the rate of formation of 2,3-DPG and its final concentration are significantly higher in affected than in normal cells and account to a large extent for the missing lactate in the affected cells.

It has recently been shown that ATP and 2,3-DPG bind to Hb under conditions that can be assumed to occur in intact cells. Addition of 2,3-DPG to concentrated hemolysates has been shown by De Verdier and Garby (27) to stimulate glycolysis, probably by competing for binding sites with ATP on Hb. Benesch and Benesch have found that 2,3-DPG exerts a profound influence on the oxygen affinity, heme-heme interaction and Bohr affect of hemoglobin solutions, facilitating oxygen transfer to tissues (29). Therefore, the changes in concentration of both ATP and 2,3-DPG found in the RBC of GSD Type III patients are expected to influence such processes. These aspects are presently under investigation.

Helpful suggestions by David Rubinstein and Rafael Gorodischer, are hereby acknowledged.

Supported in part by USPHS Grant 1 RO1 Am12672–01A1$_{HEM}$.

REFERENCES

1. RAPAPORT, S. and LUBERING, J. The formation of 2,3 diphosphoglycerate in rabbit erythrocytes. The existence of a diphosphoglycerate mutase. *J. Biol. chem.* **183**: 507, 1950.
2. SIDBURY, J. B., CORNBLATH, M., FISHER, J. and HOUSE E. Glycogen in erythrocytes of patients with glycogen storage disease. *Pediatrics* **27**: 103, 1961.
3. DANON, D. and MARIKOVSKY, Y. Determination of density distribution of red cell population. *J. Lab. clin. Med.* **64**: 668, 1964.
4. BUSCH, D. and PELZ, K. Erythrozyten-Isolierung aus Blut mit Baumwolle. *Klin. Wschr.* **44**: 983, 1966.
5. CARTWRIGHT, C. E. "Diagnostic laboratory hematology," 3rd edn. New York, Grune and Stratton, Inc., 1963, p. 42.
6. SOMOGYI, M. The determination of blood sugar. *J. biol. Chem.* **160**: 69, 1945.
7. KINGSLEY, C. R. and GETCHELE, G. Direct ultramicro glucose oxidase for the determination of glucose in biological fluids. *Clin. Chem.* **6**: 644, 1960.
8. HORN, H. D. and BURNS, F.H. Quantitative Bestimmung von L (+)-Milchsaure Dehydrogenase. *Biochim. biophys. Acta. (Amst.)* **21**: 378, 1956.
9. VALENTINE, W. N., OSKI, F. A., PAGLIA, D. E., BAUGHAN, M. A., SCHNEIDER, A. S. and NAIMAN, J. L. Hereditary hemolytic anemia with hexokinase deficiency. *New Engl. J. Med.* **271**: 1, 1967.
10. HELLER, P., WEINSTEIN, H. G., WEST, M. and ZIMMERMAN, H. J. Enzymes in anemia. *Ann. intern. Med.* **53**: 898, 1960.
11. BURNS, F. H., DUNWALD, E. and NOLTMAN, E. Über den Stoffwechsel von R-5-P in Hemolysaten. *Biochem. Z.* **330**: 497, 1958.
12. KORNBERG, A. and HORRECKER, B. L. Glucose 6 phosphate dehydrogenase, in: Colowick, S. P. and Kaplan, N. O. (Eds.), "Methods in enzymology." New York, Academic Press, Inc., 1955, v. 1, p. 323.
13. HORRECKER, B. L. and SMYRNIOTIS, P. L. 6 phosphogluconic dehydrogenase, in: Colowick, S. P. and Kaplan, N. O. (Eds.), "Methods in enzymology." New York, Academic Press, Inc., 1955, v. 1, p. 323.

14. TANAKA, K. R., VALENTINE, W. N. and MIWA, S. Pyruvate kinase deficiency in hereditary non-spherocytic hemolytic anemia. *Blood* **19**: 267, 1962.
15. HERS, H. G. α-Glucosidase deficiency in generalized glycogen storage disease (Pompe's disease). *Biochem. J.* **86**: 11, 1963.
16. CORNBLATH, M. D., STEINER, F., BRYAN, P. and KING, J. Uridine diphosphoglucose glucosyltransferase in human erythrocytes. *Clin. chim. Acta* **12**: 27, 1965.
17. LARNER, J. Branching enzyme from liver, amylo-1,4-1,6-transglucosidase, in: Colowick S. P. and Kaplan, N. O. (Eds.), "Methods in Enzymology." New York, Academic Press, Inc. 1955, v. 1, p. 222.
18. KRIMSKY, I. D-2,3-diphosphoglycerate, in: Bergmeyer, H. U. (Ed.), "Methods of enzymatic analysis." New York, Academic Press, Inc., 1965, p. 238.
19. FISKE, C. H. and SUBBAROW, Y. The colorimetric determination of phosphorus. *J. biol. Chem.* **66**: 375, 1925.
20. ADAM, H. Adenosine 5'-triphosphate in: Bergmeyer, H. U. (Ed.), "Methods of enzymatic analysis. New York, Academic Press, Inc., 1965, p. 539.
21. ADAM, H. Adenosine 5'-diphosphate in: Bergmeyer, H. U. (Ed.), "Method of enzymatic analysis." New York, Academic Press, Inc., 1965, p. 573.
22. BUCHER, T. and HOHORST, H. J. Dehydroxyaceton phosphate, fructose-1,6-diphosphate and d-glyceraldehyde-3-phosphate, in: Bergmeyer, H. U. (Ed.), "Methods of enzymatic analysis." New York, Academic Press, Inc., 1965, p. 246.
23. STEINITZ, K. Laboratory diagnosis of glycogen diseases. *Advanc. clin. Chem.* **9**: 227, 1967.
24. SOMOGYI, M. Notes on sugar determination. *J. biol. Chem.* **195**: 19, 1952.
25. JOHNSON, J. A., NASH, J. D. and FUSARO, R. M. An enzymic method for the quantitative determination of glycogen. *Ann. Biochem.* **5**: 379, 1963.
26. HOOF, VAN. -Amylo-1,6-glucosidase activity and glycogen contents of the erythrocytes of normal subjects, patients with GSD and heterozygotes. *Europ. J. Biochem.* **2**: 271, 1967.
27. DEVERDIER, C. H., GARBY, I. and HJELM, M. Intraerythrocyte regulation of tissue oxygen tension. *Acta. Soc. Med. upsalien* **74**: 209, 1969.
28. ROSE, I. A. and WARMS, J. V. B. Control of glycolysis in human red blood cell. *J. biol. Chem.* **21**: 4848, 1966.
29. BENESCH, R. and BENESCH, R. E. Reciprocal binding of O_2 and 2,3-DPG by human hemoglobin. *Proc. nat. Inst. Sci. India B.* **59**: 256, 1968.

DISCUSSION

J. C. KAPLAN (*France*): I think that it might be confusing to use the same nomenclature, e.g. A, B, ... for designating both electrophoretic bands and different types of PK deficiency, since they are not correlated. Are there clinical differences between types A and B PK deficiency? I am particularly impressed by the fact that in type B PK deficiency there is enzyme activity up to 80%. Do these patients have a severe nonspherocytic hemolytic disease? I have understood that there is no electrophoretic difference between types A and B PK deficiency. In both you found that the C band was missing. Therefore, I would like to know how you correlate this fact with the obvious kinetic differences that you found between type A and type B enzyme? What could the significance be of the three electrophoretic bands that you observed in the normal red cell?

D. BUSCH (*Germany*): The nomenclature is confusing, however, it has a historical origin. In 1967, when we described types A and B PK deficiency, we did not know about the three electrophoretic bands, A, B and C, in normal hemolysate. The diagnosis of type B PK deficiency is important. These patients have a very severe nonspherocytic hemolytic anemia and family studies showed that they were homozygotes. Both parents of these patients have low enzyme levels in their erythrocytes but are clinically healthy, while their offsprings have a very severe hemolytic anemia, in spite of an enzyme activity of 50 to 70% of normal. It was because of this that we looked for qualitative differences between the A, B and normal phenotypes of PK deficiency. The electrophoretic behavior of the enzymes in types A and B PK deficiency is the same but the genetic data are different. This is an unsolved problem and is presently under investigation. Of special interest is the fact that we cannot find the C band on the electrophoretic strip, but the corresponding fraction to type C in normal erythrocytes can be obtained from PK deficient red cells by ammonium sulfate precipitation. The latter fraction is not activated by FDP.

The last question concerns the significance of the three bands in normal erythrocytes. This problem is also under investigation. Different types of PK exist in the body. They have been shown in the brain, liver, erythrocytes and leukocytes, and are called I, II, III and types L and M in the liver. Until recently we did not know about the existence of three fractions in the erythrocytes. It is possible that type M in the liver and fraction C in the erythrocytes are the same. Their kinetic behavior is similar, being influenced allosterically by FDP and other compounds.

G. IZAK (*Israel*): I was very much intrigued by the therapeutic attempts made by you. If I understood correctly, you managed to repair most of the metabolic defects, if not all of them, in the red cells of these patients. Is that true? Why didn't you correct the life span to normal once the metabolic defect was corrected? How much inosine did you infuse and what sort of reactions have you noticed?

D. BUSCH: The last question first: The infusion of inosine was done slowly over a 3-hr period, and the patient had no remarkable reaction to it. The type B patient, S.W. (Fig. 10, p. 202), received 1mmole inosine and 0.05 mmole adenine and guanosine, each daily. Type A patient, R.Z. (Table 2 of same paper), had formerly received 10 mmole inosine, 0.5 mmole adenine and 0.25 mmole guanosine daily, also with no remarkable hypotensive response during a 3-hr i.v. infusion. Uric acid levels in blood should be measured during such infusions and the infusion should be stopped when uric acid levels rise. The patient with type B PK deficiency responded, while the type A PK-deficient patient had very little if any hematological response. The type B patient's red cell survival time improved. The metabolism of the red cells during the infusion was not quite normal. ATP level was normal for 24 hr after the infusion, but this doesn't mean that it continued to be normal.

J. MAGER (*Israel*): The assumption that the hematological disorders observed in this study are attributable to reduced PK levels, implies that under the conditions prevailing in the cell, the activity of this enzyme is much lower than that observed in the cell-free assay system. Otherwise, the PK activity would not be rate determining with respect to the overall efficiency of the glycolytic system. Would you agree with this interpretation?

D. BUSCH: Yes, I agree. Generally, in normal as well as in PK-deficient red cells, PK actual activity is very much lower than that observed in the cell-free assay system because of the strong undersaturation with its substrate PEP; the actual intracellular activity is only about 1 to 5% of the total enzyme activity. In type B PK deficiency, the intracellular activity of the enzyme is further diminished by the partial loss of the activating effect of FDP. The final metabolic defect of the PK-deficient erythrocyte seems to be indirect, being only a secondary consequence of PK deficiency. If residual enzyme activity is more than 10 to 20% of normal, the turnover of glucose to pyruvate and lactate is not rate limiting. At this level the enzyme is regulated, rather than regulating, as we have shown earlier (1965). The shortened life span of the PK-deficient red cell seems to be due to secondary effects of elevated PEP, which cause an increase of 2,3 DPG. This metabolite is an inhibitor of phosphofructokinase.

B. RAMOT (*Israel*): I don't understand the high ATP levels. Hereditary high ATP levels have been described by Brewer. Those patients are not anemic. Furthermore, Mills retracted his report on high ATP levels and hemolytic anemia since he recently found an unstable Hb sabine in this patient. I therefore raise the question of whether the finding of a disease and a biochemical abnormality in the same patient indicates any relationship between the two.

D. BUSCH: It is important to remember that the ATP level is not in itself a measure for the rate of ATP-turnover, which alone is essential for cell functions (ionic pump, etc.). In a case of "high red cell ATP level" for instance, which we described recently (*Klin. Wschr.* 1970), we could indentify a diminution of ATP-utilization-capacity of the erythrocytes as a cause of their shortened life span. In contrast to the case of Brewer, which you mentioned, and to further cases of Loos, Prins et al., our patient with high red cell ATP suffers from a severe hereditary nonspherocytic hemolytic anemia. So, in PK deficiency too, we should measure the ATP-turnover rate. And to your last question: I think, we can be sure that type B PK deficiency is also really due to a defect of the enzyme PK for all the reasons mentioned above.

16952
10/7.